Here and Now

contemporary science and technology
in museums and science centres

Here and Now

contemporary science and technology
in museums and science centres

edited by Graham Farmelo and Janet Carding

Proceedings of a conference held at the
Science Museum, London
21–23 November 1996

Science Museum with the support of the
European Commission Directorate General XII

British Library Cataloguing-in-Publication Data
A catalogue record for this publication is available from the British
Library

Set from Pagemaker in Postscript Monotype Plantin Light
Printed in England by Hobbs the Printers, Southampton, UK
Cover design: Agenda Design Associates

ISBN 0 901805 97 1

Science Museum, Exhibition Road, London SW7 2DD

Contents

List of contributors

Munkith Al-Najjar	Senior Scientist, Science North, Sudbury, Canada
Andrea Bandelli	Coordinator of Internet Development, newMetropolis, Amsterdam, the Netherlands
Jana Bennett	Head of BBC Science, London, UK
Patrick Besenval	Chief Executive Officer, Les Productions du Futuroscope, Paris, France
Sandra Bicknell	Deputy Head, National Railway Museum, York, UK
James Bradburne	Head of Design, newMetropolis, Amsterdam, the Netherlands
Mathis Brauchbar	Partner, Locher, Brauchbar and Partner, Basel, Switzerland
Robert Bud	Head of Bioscience Group, Science Museum, London, UK
Janet Carding	Wellcome Wing Project, Science Museum, London, UK
Charles Carlson	Director, Life Sciences, Exploratorium, San Francisco, US
Paul Caro	Délégué aux Affaires Scientifiques at La Cité des Sciences et de l'Industrie and Directeur de Recherche at CNRS, Paris, France
Dominique Cornuéjols	Head of Communications, European Synchrotron Radiation Facility, Grenoble, France
Sir Neil Cossons	Director, Science Museum, London, UK
John Durant	Assistant Director, Science Museum, and Professor of the Public Understanding of Science, Imperial College, London, UK
Graham Farmelo	Head of Exhibitions, Science Museum, London, UK
Wolf Peter Fehlhammer	Director, Deutsches Museum, Munich, Germany
Birte Hantke	Curator, Deutsches Hygiene-Museum, Dresden, Germany
AnnMarie Israelsson	Director, Teknikens Hus, Luleå, Sweden
Roland Jackson	Head of Education, Science Museum, London, UK
Sir Harry Kroto	Professor of Chemistry, School of Chemistry, Physics and Environmental Science, University of Sussex, Brighton, UK
David Lowenthal	Professor Emeritus of Geography, University College London, and Visiting Professor of Heritage Studies, St Mary's University College, Strawberry Hill, Twickenham, UK
Heather Mayfield	Project Director, *Challenge of Materials* gallery, Science Museum, London, UK
Arthur Molella	Director, Lemelson Center, National Museum of American History, Smithsonian Institution, Washington DC, US

Alan Morton	Curator of Modern Physics, Science Museum, London, UK
Oliver Morton	Former editor, *Wired* magazine, London, UK
Frank Olsen	Director of Public Services, Experimentarium, Hellerup, Denmark
Per-Edvin Persson	Director, Heureka, Vantaa, Finland
Melanie Quin	Director, PanTecnicon project, Techniquest, Cardiff, UK
Peggie Rimmer	Head of the Communication and Public Education Group, CERN, Geneva, Switzerland
Simon Schaffer	Reader in History and Philosophy of Science, Cambridge University, and Fellow of Darwin College, Cambridge, UK
Esther Schärer-Züblin	Project Leader, Alimentarium, Musée de l'Alimentation, Vevey, Switzerland
John Shane	Vice President: Programs, Museum of Science, Boston, US
Laurence Smaje	Director, Wellcome Centre for Medical Science, London, UK
Joan Solomon	Lecturer in Research, Department of Educational Studies, Oxford University, UK
Gillian Thomas	Chief Executive, Bristol 2000, Bristol, UK
Helena von Troil	Senior Exhibit Coordinator, Heureka, Vantaa, and Secretary for the National Committee on Biotechnology, Finland
Edward Wagner	Coordinator of the Albert M Greenfield Cutting Edge Gallery and Musser Choices Forum, Franklin Institute Science Museum, Philadelphia, US
Lorraine Ward	Exhibitions Manager, Science Museum, London, UK
David Wark	Lecturer at the Department of Physics, Oxford University, and a member of the collaborative project at the Sudbury Neutrino Observatory, Canada

Sources of images

Page 17, Robert Brook/Environmental Images, London

Pages 19, 102, CERN Photo, Geneva

Pages 21, 129, 149, 245 (fig. 1), Science Museum/Science & Society Picture Library, London

Page 24, Richard Lamouneux/Cité des Sciences et de l'Industrie, Paris

Pages 25, 53, 54, 55, 59, 60, 61, 62, BBC, London

Page 45, Deutsches Museum, Munich

Page 68, Science North, Sudbury

Pages 73, 75, Sudbury Neutrino Observatory collaboration, Sudbury

Page 78, markkinointi/Heureka, Vantaa

Pages 92, 96, 135, Smithsonian Institution, Washington DC

Page 107, Artechnique/ESRF, Grenoble

Pages 113, 120, Wellcome Centre for Medical Science, London

Page 114, Wellcome Institute Library, London

Page 121, Science Museum, London

Pages 140, 143, Technikens Hus, Luleå

Page 152, Royal Society, London

Page 157, Adam F Kelly–Vincent J Massa/Franklin Institute Science Museum, Philadelphia

Page 175, Diversity University

Page 181, Cees van Giessen/CIIID, Amsterdam

Page 216, H Vimenet/Futuroscope, Paris

Page 221, Arnaud Legrain/Cité des Sciences et de l'Industrie, Paris

Page 237, Popperfoto/Reuters, London

Page 245 (fig. 2), Zenyx Scientific, Manchester

Pages 251, 282, Heureka, Vantaa

Page 259, Locher, Brauchbar and Partner, Basel

Page 267, Alimentarium, Vevey

Page 269, Deutsches Hygiene-Museum, Dresden

Pages 274, 276, Joe Samberg Photography/Exploratorium, San Francisco

All other images supplied by the editors.

Preface

Sir Neil Cossons

The twentieth century has arguably been the scientific century. There have been a remarkable number of developments in science and technology over the past hundred years, and it is surely no coincidence that they have occurred at the same time as a golden age for science museums and the birth of the science centre.

While many science museums and science centres have flourished, all of them have found it difficult to respond to the speed of change in science, technology and medicine. Typically, the public still find out about contemporary science and technology mainly from media such as television, radio and newspapers rather than from museums and science centres. Yet why should this be so? Is there a specific role that science museums and science centres can play by providing access to authoritative information and to the scientific debate?

In recent years, many museums and science centres have made great efforts to address contemporary issues. Here at the Science Museum, we are building on the success of our *Science Box* and *Technology Futures* exhibition series to develop an extension to the Museum, the Wellcome Wing, to be devoted in its entirety to contemporary biomedical science and information technology.

The issues surrounding the presentation of contemporary science and technology were comprehensively addressed at the conference, *Here and Now*, which we organised at the Science Museum from 21 to 23 November 1996, as part of the fourth European Week for Scientific and Technological Culture. This event, generously supported by the European Commission Directorate General XII, was the first conference to be devoted to the challenge of improving the presentation of contemporary science and technology in museums and science centres.

Among the 184 delegates from 18 countries were not only leading members of the museum and science centre community, but also scientists from industry and universities, historians and media figures, together with representatives from major European research laboratories and funding organisations. This remarkably eclectic gathering enjoyed a rich selection of talks and discussions addressing the challenge of presenting contemporary science from a wide variety of perspectives.

Each of the papers in this collection is based on a presentation at the *Here and Now* conference. In keeping with the breadth of interests represented among the delegates, the range of contributions is extraordinarily diverse.

The 35 papers presented here give, we believe, an insight into the challenges that museums and science centres face as they increase their coverage of contemporary issues. The challenge for the next millennium is to do more than reflect on past achievements of scientists and technologists by improving our coverage of the work they are doing today. In meeting this challenge, we shall be truer to the spirit of scientific and technological enterprise and be discharging our responsibility to preserve material culture. Moreover, our great historic collections will be seen in the context of the present.

We hope this book will serve to widen the debate, begun at the *Here and Now* conference, about the role of museums and science centres in the twenty-first century.

Sir Neil Cossons is Director of the Science Museum, London

Foreword

Sir Harry Kroto

As we approach the twenty-first century, scientists are one of the few groups of people with anything like a clear view of the profound problems that must be solved if the human race is to survive. It is crucial, if for no other reason than this, that the importance of science and technology in today's society is understood more widely, particularly by young people. Museums and science centres have a potentially key role to play in this, especially in the way they present contemporary science and technology.

Communities in the third world that have not been able to take advantage of the advances in science and technology have lifestyles roughly equivalent to those of the European fifteenth century or earlier. However, the material standard of living of the 'advanced' nations has been achieved at enormous ecological and environmental cost, involving the consumption of huge amounts of energy and depending critically on efficient farming methods, which also involve huge energy costs.

Despite the efforts of our museums and science centres, it is also clear that the majority of people in the first world have become oblivious to the fact that they are totally dependent on the fruits of science and technology. They switch on electrical appliances, consume high-quality food (heated or cooled to order), wear clothes made from synthetic materials, live in accommodation constructed from modern materials and travel around the world with scarcely a thought of what an amazing environment has been created.

The level of ignorance in some parts of the media about the role of science in society is perpetually astonishing. No less a figure than Simon Jenkins, the influential former editor of *The Times*, London, wrote in 1994: 'The national curriculum [for schools] puts quite unrealistic emphasis on science and mathematics which few of us ever need.' With this kind of uninformed statement having currency in the UK among opinion formers, it is hardly surprising that the role of science is so seriously underestimated. Moreover, anti-science attitudes are becoming worryingly prevalent in the rest of Europe, the US and elsewhere.

In my view, a better understanding of today's science and technology is essential if the public is to be able to influence positively the way our society progresses in the future. Science museums and science centres form a key element in making the public aware of just how precariously balanced our lifestyle is. These institutions must meet this challenge by inspiring our youngsters to consider science, engineering and technology not just as vital to our survival with a reasonable standard of living but also as cultural activities that can provide as much intellectual stimulus and excitement as any other subject.

It is particularly important that a balanced presentation of science and technology is achieved. For instance, basic engineering skills are as valuable and exciting as the most modern high-tech computational skills. While it is important to recognise the seductive charms of modern technology, they must be balanced with thoughtful presentations of the elegance of the more basic—and older—technologies which everyday familiarity has caused most of us to take for granted. Museums and science centres must continue to play a major role in showing how such old technologies were as revolutionary in their day as the microchip is now. But they also have a key role to play in encouraging their visitors to engage with the issues of modern science and technology.

Unless the young people of the twenty-first century appreciate the importance of science, we stand no chance whatsoever of economic, social or cultural survival. In my view, science museums and science centres must play an appropriately active part in the educational programme on which this survival depends.

Sir Harry Kroto was joint-winner of the Nobel prize for chemistry, 1996

Introduction

Graham Farmelo and Janet Carding

How can museums and science centres improve their presentation of contemporary science and technology? The papers in this collection shed light on the issues and suggest ways ahead.

Every science museum finds it hard to keep up with science. This is scarcely surprising when you bear in mind that most museums—like their sibling science centres—only have the resources to develop slowly, if at all, whereas science and technology are continually developing apace.

Unable to stay abreast of contemporary science,[1] museums and science centres struggle to cope with the welter of controversy, complexity and uncertainty at its frontiers. As a result, visitors have to make do with displays that rarely do justice to the latest thinking and that are more often than not at least a decade out of date.[2]

Yet there are some encouraging signs of change as museums and science centres seem to be becoming less and less willing to accommodate tired old exhibitions, while they are more and more keen to tackle the latest stories. In this collection of essays, the problems that museums and science centres have to solve if they are to keep up to date are discussed by experts who work in these institutions and by colleagues in the related fields of the history of science, the media and scientific research.

A wide range of questions is addressed from many points of view. How effective is the presentation of 'live' events as a way of responding to the latest science (figure 1)? What policies should museums adopt if they are to use their collections in contemporary displays? Will the prospective virtual museum visit render the traditional type of visit obsolete? Do visitors, especially the all-important young cohort, actually want to know about modern science, or would they be just as happy reflecting on the past?

The pieces here range widely across topics in science, but with a final section that focuses specifically on biotechnology, the rapidly emerging field that promises to be one of the most important branches of industry in the next millennium. What can we learn from previous experiences in this discipline, which is bound to figure more and more

Figure 1. Are lectures, workshops and demonstrations the best way of introducing contemporary science into museums and science centres?

prominently in the public mind in the coming decades?

The collection is based on the *Here and Now* conference held at the Science Museum, London, on 21–23 November 1996.[3] This conference brought together 184 delegates from a wide variety of fields to discuss the challenges of presenting contemporary science and technology in museums and science centres. The first three essays are based on the keynote addresses, given by an eminent historian, by the director of one of the world's leading science museums and by a leading television programme maker. These pieces address topics that are at the heart of the debate: whether museums should even try to keep up with science and technology, the slippery meaning of 'contemporary' and the alleged tendency of the media to give utopian treatments to contemporary scientific topics.

It is important to bear in mind that the word 'contemporary' is not easy to define. Simon Schaffer, the distinguished historian of science from the University of Cambridge, perceptively analyses the difficulties associated with the display of contemporary scientific developments (p31). He shows that when scientists are debating the 'facts', museums must be acutely aware of institutional affiliations, previous arguments and the history of the current controversy. He suggests an ambitious programme in which this information is used to produce open-ended and opinionated exhibitions that juxtapose different perspectives from a variety of sources. In order to present contemporary science effectively in museums and science centres, he asserts, we need a historical understanding of the meaning of 'contemporary', along with an awareness of the often tortuous paths taken by the sciences as they develop.

Wolf Peter Fehlhammer, Director of the famously conservative Deutsches Museum in Munich, argues that museums have no choice but to keep up with science if they are to fulfil their role of communicating with the public about the true nature of science (p41). Fehlhammer's own museum is now fashioning

plans for an extension in *Vision 2003*, which will considerably increase contemporary coverage and, he suggests, bring the museum into line with its original purpose as stated by its founders almost a century ago. It might seem that such a transformation would be relatively easy for an institution as well resourced as the mighty Deutsches Museum, but actually it is much harder to change the direction of a huge institution like this—especially bearing in mind its responsibilities to preserve its collections—than it would be to build a new museum from scratch.[4]

Much of the contemporary science in the broadcast media comes from the deadline-driven world of news reporting. This is perhaps one reason why so much of this news and, to a lesser extent, feature programmes, are often criticised as being utopian, with the accusation that they tend to present a sanitised view of science and the scientists.[5]

Jana Bennett, Head of the Science Unit at the BBC, believes that although this might have been true 30 years ago about contemporary science reporting on BBC television, programmes are now much more critical of science (p51). She believes that, rather than promoting the importance of science-for-science's sake, today's programmes on science in the news must concentrate on showing how the latest developments will affect people's daily lives. In museums, too, exhibition and programme developers are having to adopt a much more critical, less starry-eyed approach to scientific and technological issues.

Programming for success

If the gallery is the museological counterpart of a television programme, then temporary exhibitions and events (often collectively known in museums and science centres as 'programmes') are the counterparts of news bulletins. A gallery is normally several years in the making, whereas a modest exhibition or a demonstration can be put together in weeks or sometimes days. If we regard 'contemporary science' as science that has only been in the public domain for a few months (admittedly a rather hard-line and naive definition), then the most natural medium for its presentation is programming.

It is through an analysis of its programmes that you can usually see how seriously a museum or science centre takes contemporary science. In 1996, when the putative link between bovine spongiform encephalopathy (BSE or 'mad cow disease') and a variant of its human counterpart, Creutzfeldt–Jakob disease (CJD),[6] was the hottest science news topic in the UK, not one of its major science museums and science centres had presented anything on the subject within three months of the story's breaking in March (figure 2).[7] Philip Campbell, the editor of *Nature*, was by no means the only speaker at the *Here and Now* conference to lament this and he urged museums to cover topics as difficult as this.[8]

Less emotive than BSE, but no less intellectually challenging, is the physics of neutrinos, a topic addressed in an extraordinarily ambitious set of programmes at Science North in Canada. Munkith Al-Najjar describes how this science centre has worked with international teams of scientists at the nearby Sudbury Neutrino Observatory to bring this recondite area of sub-atomic physics to visitors through a variety of exhibitions, special events and activities (p67). David Wark, one of the physicists working at the Observatory, describes the underlying science and gives his impressions of this collaboration from his point of a view as a scientist prepared to meet the challenge of communicating with the public (p73).

During the past decade or so, drama has become increasingly popular as an interpretive technique. It is not unusual to walk into a museum or science centre to see an actor performing in a role that sheds light on an exhibition or theme. As popular as this technique has proved to be, especially with younger visitors, it has not always proved easy to adapt it to deal with contemporary topics.[9]

Melanie Quin of the UK's Techniquest science centre in Cardiff suggests that

role-play, in the form of street theatre, can be an effective way of engaging the public inside science centres and also outside, via outreach activities (p77).[10] Encounters with real scientists, too, are often popular with visitors, provided the atmosphere is agreeable and if the scientist is a good communicator.[11] For Quin, the mark of a good science centre is its ability to engage in a dialogue with its visitors and she argues that this is best achieved by using more people-mediated activities and less exhibition hardware.

Figure 2. The possible link between BSE and CJD was one of the hottest science news stories of 1996, yet most museums and science centres did not address it

For most museums and science centres, events are too labour intensive (and, therefore, too expensive) to be the main part of the diet they offer their visitors—it is through exhibitions that most of their contemporary science content is presented. An important challenge is to find ways of responding to new developments at short notice and to secure sufficient resources to be able to present a continuous series of exhibitions.

Lorraine Ward describes how a three-year sponsorship agreement made this possible at the Science Museum, London (p83). The series, known as *Science Box*, comprised 13 small exhibitions that each addressed a topic in contemporary science, for example, chaos, nanotechnology, infertility treatment or the information superhighway. Several of these exhibitions toured the UK to venues such as libraries, hospitals and shopping centres, most of which would otherwise feature little or no contemporary science.

The idea of touring contemporary science exhibitions is gradually becoming more popular in Europe and the US, no doubt mainly because it makes such good economic sense. As yet, however, remarkably few contemporary science exhibitions travel across national boundaries and there is plenty of room to expand programmes like this. In order to achieve this, it is vital to encourage the spirit of international collaboration through organisations such as the European Collaborative for Science, Industry and Technology Exhibitions (ECSITE) and the Association of Science–Technology Centres (ASTC).

A notable recent development in contemporary programming in the US was the opening of the Lemelson Center for the Study of Invention and Innovation, an exciting new focus for programming activities in the National Museum of American History, part of the Smithsonian Institution in Washington DC. The Center aims to record the post-war past, to broaden the public understanding of invention through a wide range of programmes, and to encourage young people to study and explore the idea of invention. The director of the Center, Arthur Molella, describes some of its activities, suggesting that the importance of technology and invention are made easier to appreciate by broadening their contexts to include the arts and other fields (p91). To underline the point, one of the most successful events the Center has organised included a symposium and exhibition focusing on an icon not of engineering but of the post-war entertainment industry: the electric guitar.

Collaborate or stagnate?

One of the most obvious ways in which museums and science centres can improve their contemporary coverage is by joining forces with other institutions, whether it be with academic organisations, sponsorship or funding agencies, or with industrial partners. In this section, authors from five countries describe a wide variety of collaborations that they have forged with other institutions. The first two papers concern two of the large European research facilities, the European Laboratory for Particle Physics (CERN, figure 3) in Geneva, Switzerland, and the European Synchrotron Radiation Facility (ESRF) in Grenoble, France.

Peggie Rimmer, Head of the Communication and Public Education Group at CERN, and Heather Mayfield of the Science Museum, London, describe how they have worked together, in order to give each institution the benefit of the other's strength (p101). While the Science Museum was able to provide advice on the best way of developing CERN's exhibition policy, CERN was able to provide expert advice during the production of the Museum's 'Big Bang' exhibition, which was presented in CERN for five weeks in early 1997, after its run at the Museum.

Peggie Rimmer's counterpart at the European Synchrotron Radiation Facility, Dominique Cornuéjols, reports that her attempts to foster collaborations with museums and science centres have been rather less fruitful than CERN's (p107). The most successful collaborative ventures that the ESRF has undertaken have been within the framework of the European Weeks for Scientific and Technological Culture, organised under the aegis of the European Commission's Directorate General XII.[12] Some aspects of the account given here by Cornuéjols will not make happy reading for those of us in European museums and science centres: the hand of collaboration has been offered to us and we have apparently turned away.

Figure 3. Left: one of the huge particle detectors at CERN, the European atom-smasher. How can museums preserve technology on this scale?

Our spirits are lifted by Laurence Smaje's account of a bevy of successful collaborative projects funded by the Wellcome Trust, the world's largest medical research charity, based in London (p113). In 1991, the Trust established the Wellcome Centre for Medical Science in order to explain the nature of medical research to the public. The Centre undertakes activities of its own but, as Smaje notes, it has also found it beneficial to fund commissioned work and research submitted by others. Smaje reviews the benefits of collaboration, as well as the attendant difficulties, and he describes several case studies that illustrate his belief that partnership thrives in a climate of mutual trust. For example, the Trust has—after many years of successfully collaborating with the Science Museum, London—contributed substantially towards the building of its Wellcome Wing, which will focus on contemporary biomedical science and information technology. In another example, the Trust joined forces with the UK's Y-Touring theatre company in the development and touring of *The Gift*, a unique play that addresses some of the difficult issues of modern genetics. Thanks partly to a powerful yet sensitive script by the playwright Nicola Baldwin, the production proved very popular with audiences all over the UK, notably teenagers.

Working with industry can sometimes be challenging for museums and science centres, always jealous of their editorial independence. Yet, as Gillian Thomas explains, it is possible— given sufficient planning and attention to detail—to set up sponsorship arrangements that benefit both parties (p125). She illustrates the breadth of opportunities that museums can offer potential sponsors, such as association with a prestigious organisation, privileged access to specific audiences and facilities that can be used outside opening hours for commercial purposes. Contemporary science displays pose special challenges for fund raisers as it is important for them to win resources not only to finance an initial exhibition but also to keep it fresh for years after its opening.

Despite the best intentions when sponsorship agreements are being set up, things do not always go according to plan. The *Science and American Life* exhibition at the National Museum of American History[13] has been one of the most talked-about examples in recent museological history of an exhibition sponsorship arrangement that went awry. The American Chemical Society, which sponsored the exhibition, were not best pleased with its developers' selection of topics and with its treatment of the social dimensions of modern science.[14] These were not the only critics, and the exhibition opened in 1994 in a storm of controversy. Arthur Molella, one of those responsible for developing the exhibition, tells the story candidly and with some sadness at the treatment meted out to the project (p131).

He believes that the Museum became unwittingly caught up in the undeclared 'science wars' between scientists and the cultural-studies community:[15] if the exhibition had opened in the 1980s or even in the early 1990s, he suggests, the story would have been very different. So long as museums and science centres depend on private funding, he believes, they will be vulnerable to outside pressure, which inevitably worsens in stressful times. He ends on a rueful note: 'at least for the near future, before we venture too far outdoors, we'd better check the weather'.

Although she has had some difficulties, AnnMarie Israelsson has had a rather happier time in dealing with her funders at the Teknikens Hus science centre in northern Sweden (p139). She describes how the centre has grown in no small measure as a result of its policy of fostering sponsorship arrangements with local firms in order to support a wide range of activities. For Israelsson, the cultivation of collaborations does not only mean gaining access to funding, it can also open up new possibilities and offer a new relationship with education and society. Other contributors to the conference supported this view; for example Richard Piani, Délégué Général of La Cité des Sciences et de l'Industrie, described how his institution has successfully developed long-term strategies for collaborating closely with both large and small industrial companies.[16]

Collections

For museums, one of today's most pressing challenges is to keep a material record of scientific and technological activity.[17] What makes this challenge so daunting is that this activity is growing exponentially, while the resources for acquisition and preservation[18] are at best stable and at worst sharply decreasing. In addition, the undisputed popularity of 'hands-on' exhibits continues to put pressure on curators to display only the most appealing items in their collections, while the rest of their treasures lie in store. There is even doubt about the capacity of museums to have sufficient room for objects that are representative of some of the 'big science'[19] experiments characteristic of our age.

Alan Morton insists that the presentation of objects in contemporary science exhibitions can add enormously to the range of stories that exhibit developers can develop (p147). For him, the way ahead is to produce exhibitions comprising a mixture of interactives and judiciously chosen objects, each adding to the power of the other. As he stresses, objects have different meanings in different contexts, so the choice of which objects to display in a given context is never self-evident, rather it is a matter of choice for the exhibition developers (figure 4).

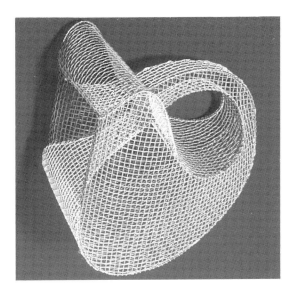

Figure 4. Material artefacts can sometimes be an important record of scientific work. These Klein bottles can be described as Mobius-strip-like solids and have only recently been understood mathematically.

The Cutting Edge Gallery at the Franklin Institute Science Museum, Philadelphia, has proved to be a popular forum for the presentation of modern technological artefacts. Edward Wagner describes how he and his colleagues have found innovative ways of inexpensively acquiring items of new technology—high-definition televisions, videophones, digital film editing equipment, etc.—before they become a familiar part of everyday life (p155). This is no small achievement: most industrial firms are loath to give (or even to loan) their latest equipment to museums, unless they perceive it to be clearly to their economic advantage. Wagner reports that demonstrations of this equipment have now become a popular feature at the Gallery, proving once again the visitors' appetite for finding out about new and futuristic technology in science museums.

Yet a central question remains: how can museums satisfy this appetite in the face of competition from other, apparently more flexible, media, such as television, radio and newspapers? The museums have the oft-stated advantage of authenticity, but only if they can be quick enough to put contemporary authentic material on display in ways that the public want to see.

But do museums overestimate the importance of objects?[20] The historian and geographer David Lowenthal has long been sceptical of the collection of contemporary objects for its own sake and has encouraged a more thoughtful approach to the collection and display of modern artefacts.[21] Here, he suggests that the key issue is not what things to collect or conserve, but how to address the concerns of visitors, for whom scientific research is increasingly invisible and who seem to be more and more fearful of its findings (p163).

Lowenthal believes that museums have to attend carefully to the lessons of the recent debates about the nature of science, particularly as advanced by those who are allegedly 'anti-science'.[22] Few would deny that these issues raised by the critics of science have profoundly influenced the public perception of science[23] and that they have made some scientists pause to think more deeply about their work. Yet this change in intellectual climate has barely been perceptible in most museums except in ways that have muffled their voices, as Arthur Molella notes (p131). Is it too much to ask for a museum to respond constructively in its exhibitions to such issues? And is there any realistic possibility that objects might play a part in such exhibitions?[24]

Virtual visit

'With the increasing pervasiveness of communication between computers, we are in the middle of the most transforming event since the capture of fire,' according to John Perry Barlow.[25] One does not have to share Barlow's views on the impact of the Internet and of other aspects of the information technology revolution, to agree that they will have a profound and potentially transforming effect on every museum and science centre in the

Figure 5. What will be the long-term impact on museums and science centres of the revolution in information technology?

world. Not only will the hardware and software at our disposal be changing, the nature of visitors will change too.[26] Although many manifestations of new information technology have emerged more slowly and less effectively than their most zealous advocates have forecast, there is no doubt that they will have a profound impact on museums and science centres over the coming decade. As Anthony Smith has noted, 'we are entering a new realm, some might say era, of high-definition, interactive and mutually convergent technologies and communication.'[27]

The 'Virtual Visit' session at the *Here and Now* conference was chaired by Oliver Morton, then the editor of the UK edition of *Wired* magazine. In an overview of the issues, he suggests that, contrary to the oft-stated view, the digital revolution is not about speed or processing power, rather it is about information and the new possibilities of interacting with it (p169).[28] As such, he hints that the revolution may be a fine opportunity for museums to reinvent themselves (figure 5).

Roland Jackson gives an idea of how this might be achieved (p173). He points out that museums can now move away from the didactic, top-down model of the museum towards a bottom-up model, in which visitors—on-line and those who physically go to the museum—not only have access to its material but also have a share in its creation. Through the welter of new Internet-based facilities such as videoconferencing, e-mail conferencing, Web-page creation and so on, participants will be able to construct their own on-line exhibitions, to discuss topical issues and to debate with experts. This is a bold and challenging vision of a new type of museum, built partly in real space and partly in cyberspace: does anyone have the imagination, expertise and courage to bring such ideas to fruition?

In the Netherlands, Andrea Bandelli and James Bradburne are certainly making a start. They describe how they are using Internet facilities to present a wide range of new programmes for their visitors at the newMetropolis science centre in Amsterdam (p181). Bandelli and Bradburne have found imaginative ways of using Internet facilities to give their visitors what are, in essence, new ways of achieving the old-fashioned virtues of the museum and science centre: social interaction in a public setting. Echoing the aspirations of Oliver Morton, they aim 'to transform visitors into users', for example by providing them with information on the latest science news and by enabling them to create information for others to use, making their institution both a platform and a forum for public opinion.

It would not be right to give a platform only to the cyberphiles. In the final piece in this section, John Shane of the Museum of Science in Boston argues the sceptic's case (p189). He asks whether the new-fangled paraphernalia of information technology really does add to the quality of visitors' experience. What proportion of visitors actually can and want to use this new technology in a museum? And can museums and science centres afford the true cost of keeping up with the latest in presentational technology? After a candid and provocative analysis, Shane concludes that the benefits of the virtual visit have been grossly exaggerated: perhaps the benefits are as virtual as the visit itself?

It may be, however, that museums and science centres in the latter days of the information age will diverge even more sharply from the current model than we can now conceive. According to the most radical agenda, 'moving museums into cyberspace means remembering what they were originally meant for, appreciating what they are actually used for, and then creating something completely different'.[29]

Visitor view

At the *Here and Now* conference, the neuroscientist Susan Greenfield suggested that most people's education in science suffers from a surfeit of facts, leaving too little room for individual expression. She added: 'the exciting element that can only come when [something] has just been discovered, or is still controversial, is frequently sadly lacking.'[30] Greenfield believes that museums and science centres are an ideal place to rekindle the public's enthusiasm by involving them in the latest science.

But do the public really want to know about contemporary science?[31] 'No' is the pithy and controversial conclusion reached by Frank Olsen (p199), although he does point out that the answer changes to 'yes', if 'the subject relates to visitors' everyday lives and/or if they have the chance to interact with museum staff'. Olsen's conclusion is based on visitor research he has carried out at the Experimentarium in Copenhagen, where he studied the responses of visitors to their 'Demos on Wheels'. These results support the widespread view that we cannot take for granted the visitors' enthusiasm for contemporary science—we have to make our coverage appealing by building on links to what is important and engaging to our visitors, rule number one of science communication.[32]

Attendance figures for museums and science centres clearly demonstrate that many of them do not have the breadth of appeal that they would like. There is, for example, a clear disparity between their appeal to males and to females. After a review of the disparate literature on this topic, Sandra Bicknell suggests strategies for redressing the balance (p205). She advocates more imaginative choices of exhibition themes, enabling museums and science centres to cater for a broader range of interests, attitudes and motivations. In the coming decades, the need for such thinking about the needs of new audiences is going to be increasingly acute if, as many experts confidently expect, the average age of visitors increases[33] and the public have more leisure time.[34]

It is a widely accepted truth in the museum and science centre community that, if you want to create a popular exhibition with broad appeal, you must feature well-designed 'interactives' or, in other words, hands-on exhibits. Although the appeal of interactivity in museums and science centres has been well documented,[35] it is also true and paradoxical that the most popular medium in the world, television, is enjoyed almost entirely passively. Might there be other ways of engaging the public in contemporary science? The film producer Patrick Besenval draws on his experience at Futuroscope near Poitiers and at La Cité des Sciences et de l'Industrie in Paris to suggest that it is visitor involvement that is crucial, not just interactivity (p215). He points to the great success of white-knuckle rides and also to the large-screen film theatres (IMAX and OMNIMAX) at Poitiers and at numerous other science centres and museums. Besenval suggests that it is important to engage 'the body, the emotions and the brain' of the visitor in order to make the maximum impact: perhaps new mass-participation, whole-body experiences like this will become an increasingly popular way of communicating contemporary science?

Such experiences would certainly be popular with school visitors. Museums and science centres have long provided popular venues for them, but it is not easy to see how these trips can contribute to students' knowledge and appreciation of contemporary science: most school curricula focus on material that is well established and therefore comprehensively written up in text books.[36]

In a wide-ranging paper, Paul Caro of La Cité des Sciences et de l'Industrie addresses the challenge of formulating an education policy for schools in a modern science centre (p219). Currently, he points out, La Cité is pursuing a 'postmodern strategy' in which they do not expect children to master vast amounts of information during each visit, but rather fragments of knowledge and methods for linking them together. He speculates that in future it might be more appropriate for his and other science centres and museums to develop this strategy so that they become resource centres supporting displays of the basic topics underlying contemporary science (figure 6).

Every museum and science centre has its own education policy, partly because of the wishes of its staff but mainly because it has to respond sensitively to the needs and wants of its local schools. Joan Solomon draws on her experience of working on the recent 'School science and the future of scientific culture in Europe' project[37] to summarise the myriad cultural difference between the educational philosophies in European countries (p227).[38] From this paper, it is plain that any museum or science centre that attempts to cater for a pan-European clientele cannot expect their task to be easy.

Figure 6. Fast-response temporary exhibitions on contemporary science have been pioneered at La Cité des Sciences et de l'Industrie, Paris

Figure 7. How would visitors to museums react to controversial biotechnology exhibits, such as a human ear growing on to the back of a laboratory mouse?

Focusing on biotechnology

The final essays in the collection all concern biotechnology, one of the defining new technologies of our age.[39] In common with all emergent technologies, its economic potential excites industrialists and its scientific subtleties intrigue its practitioners, while the perennially wary public at large look on nervously as they are largely disenfranchised from most of the debate about its implementation.[40] There is something particularly sensitive about biotechnology, for it concerns the manipulation of life itself, something that many of us intuitively feel is sacred and should therefore be left well alone (figure 7).

The distinguished biotechnologist Hans Lehrach, of the Max Planck Institute in Berlin, spoke at the *Here and Now* conference about the public fears of biotechnology. He said, in some exasperation, that he keeps telling the people of Germany that if they expend their energy worrying about dinosaurs being regenerated from DNA when their chances of being hit by a bus are much greater, they are not dealing entirely rationally with their situation.[41]

How, then, should museums and science centres approach the presentation of biotechnology? John Durant suggests that, quite apart from the usual problem of interpreting subject matter as complex and as rapidly evolving as this, exhibit developers have difficulties in deciding which of the myriad possible narrative stories to tell in order to illuminate the subject (p235). The choice hinges on whether the intention is to arouse the visitors' curiosity, to excite their admiration or to provoke them.

Robert Bud, curator and historian of biotechnology, argues that original objects are a powerful way of presenting the complex and emotive issues of the subject (p241).

Well-chosen artefacts can be extraordinarily suggestive and illuminating: for example, Alexander Fleming's Petri-dish, emblematic of the UK's supposed inability to capitalise on its fundamental discoveries, and the sexually determined mouse developed by Peter Goodfellow and Robin Lovell-Badge, the first animal to have its gender changed by scientists. Bud proposes a systemic solution to the challenge of presenting biotechnology in museums: both interactives and original objects are needed to realise the potential of displays to juxtapose the dreams and realities associated with this branch of technology, as well as to address the visitors' concerns about its impact on their lives.

The remaining essays in this section concern biotechnology exhibitions that have been presented or are in preparation in Europe and the US. Helena von Troil begins by describing her experiences at the Heureka science centre in Finland, where she and her colleagues have developed several biotechnology exhibits since 1993 (p249). Many of these exhibits have been successful with visitors, notably the model of the 'Pigcow'—a hypothetical cross between a pig and a cow—which, however, some scientists thought was too far-fetched and unnecessarily alarming. Commercial organisations, especially their marketing departments, have usually been unwilling to become involved in presenting subjects that might give rise to damaging publicity. As von Troil points out, in presenting topics as controversial as contemporary biotechnology, it is extremely difficult to keep all the interested parties happy.

Mindful of these potential difficulties, Mathis Brauchbar in Switzerland set up a special advisory panel before he began work on his biotechnology exhibition (p257). The panel comprised representatives of the stakeholders, including the government, independent environmental protection agencies, pharmaceutical companies and the public, to advise him and his colleagues on the content of the exhibition. As he describes, this was a fruitful approach and it led to the production

of a popular and widely lauded exhibition that has successfully toured Swiss cities since it opened in 1993.

A different approach has been used by the five museums (four German and one Swiss) involved in the *Gene Worlds* collaboration. Birte Hantke and Esther Schärer-Züblin describe how the participating institutions feel that museums are the ideal medium to present biotechnology because they are usually regarded as impartial by visitors (p263). The institutions agreed to explore common themes and to develop them independently, calling on each others' expertise when necessary. They also agreed to the same 1998 opening date and to share publicity, a catalogue and complementary events.

One of the most ambitious exhibitions to cover biotechnology-related material to date is *Diving into the Gene Pool*, presented in 1995 at the Exploratorium in San Francisco. This commercially supported venture comprised 28 exhibits, took four years to develop and attracted some 180,000 of their 274,000 visitors during its five-month run. Charles Carlson concludes that the exhibition was a qualified success (p271). Survey data for the exhibition demonstrated that the technical concepts in genetics and biotechnology that the developers presented were beyond the common experience of most of the visitors, although many of the social and ethical issues proved particularly successful in engaging their attention.

One of the most sobering figures that Carlson reports is the average time spent by the Exploratorium's visitors at the exhibition: eight minutes. This is typical of the amount of the visitors' time that exhibit developers have in order to get their message across, and it underlines the need for them to be selective when they are choosing which biotechnological stories to tell.

Summing up

What, then, are the conclusions? How *can* museums and science centres improve their presentation of contemporary science and technology? The answer appears to lie in collaboration. Through joint ventures and by working with colleagues in other fields, museums and science centres can develop their potential to offer their visitors an authentic experience, through real things, real people and real phenomena. Trying to do this in the midst of the development of potentially transforming new information technologies will be a formidable challenge.[42]

No museum or science centre can hope to cover all the facets of contemporary science by 'going it alone'. There is an urgent need to forge new and imaginative partnerships that will enable museums and science centres to tap into scientific expertise, to share interpretive skills and to gain the financial resources that will enable continual renewal. Among scientists, currently more aware than ever of the need for good public relations, there is a new willingness to share knowledge and to have a dialogue with the public. With their experience of communicating with the public, experts in the fast-response media of radio, television and newspapers have much to teach their colleagues in museums and science centres about responding swiftly to science news, about thoughtfully covering longer-term developments and about making potentially abstruse stories palatable to the public.[43]

There is no avoiding the fact that covering contemporary science, in any medium, is expensive. Responding to news items from the previous few months, let alone the previous days, is greedy of researchers' time and demands huge refurbishment budgets for exhibitions. If a museum or science centre wishes to give even a semblance of adequate coverage to modern science, then it ultimately has no choice but to win new funds or to change the priorities it gives to its other activities.

Many museums and science centres are indeed having continually to rethink their role in the increasingly competitive marketplace of leisure and education. Not least among the factors causing these constant re-evaluations is the continuing revolution in information technology. The currently burgeoning Internet is bringing a wealth of new opportunities that foreshadow potentially dramatic developments in virtual reality.[44] This technology will in principle enable its users to reveal the invisible, to conjure the unseeable, to realise environments that can now only be dreamt of. As the writer J G Ballard has suggested, 'If virtual reality comes on stream, it will represent the greatest challenge to the human race since the invention of language.'[45] What, in this technological climate, will be the value of the much-vaunted authentic experience offered by museums and science centres?

Authenticity is, however, not their only important attribute; they also offer a unique communal experience, the only way in which most visitors can find out about science in the company of others. Over and over again, visitor surveys have demonstrated the importance of this public dimension and of making the environment congenial, comfortable and non-threatening.[46] As museums and science centres strive ever harder to reach remote-access visitors, they cannot afford to neglect the need to give their visitors a traditionally warm welcome.

It is not surprising that these thoughts chime with the conclusions of the *Here and Now* conference rapporteur, Per-Edvin Persson (p281), who surmises that 'museums and science centres will become less dependent on location, but more dependent on people'. Although modern communications technology has drastically undermined the importance of distance, people still have a vital part to play in all the media involved in trying to engage the public in contemporary science—museums and science centres are no exception.

Notes and references

1 From here on in this article, we shall use 'science' to refer collectively to both science and technology, except where we specify otherwise.

2 For interesting personal views of some of the world's leading science museums, see 'Ten great science museums', *Discover*, 14 (1993), pp77–113.

3 Turney, J, 'What button do you press to turn on tomorrow?', *The Times Higher Educational Supplement* (20 December 1996), p17

4 Garfield, D, 'The next thing now: designing the twenty-first century museum', *Museum News*, 75 (1996)

5 Newcomb, H, *Television: The Critical View*, 4th edn (New York: Oxford University Press, 1987); see also de Cheveigné, S, and Véron, E, 'Science on TV: forms and reception of science programmes on French television', *Public Understanding of Science*, 5 (1996), pp231–53.

6 Will, R, Ironside, J, Zeidler, M, *et al.*, 'A new variant of Creutzfeldt–Jakob disease in the UK', *Lancet*, 347 (1996), pp921–25; see also the statement by the UK Spongiform Encephalopathy Advisory Committee, House of Commons, 20 March 1996 (London: HMSO).

7 This is the result of a telephone survey of 16 leading science museums and science centres in the UK, carried out by the Science Museum, London, in the first week of June 1996.

8 Campbell, P, presentation to the *Here and Now* conference (London: Science Museum, 21 November 1996)

9 During the DNA fingerprinting exhibition presented by the Science Museum, London, in 1992, the Museum's resident drama company created a character of a press photographer, who talked about the role of the fingerprinting technique in a (fictitious) assignment and about other (actual) cases. When playing this role, the actor was dressed in modern-day clothing, so that he or she resembled any other adult. Mainly as a result of this, visitors were nonplussed by what they initially perceived as the antics of a somewhat eccentric fellow member of the public. When the audience realised that the performer was an actor, many of them felt that they had been deceived. The actors who played the role were grimly aware of the unpopularity of this performance and some were reluctant to play the role. By common agreement, the role was rapidly withdrawn.

10 Quinn, S, and Bedworth, J, 'Science theatre: an effective interpretive technique in museums', *Communicating Science to the Public* (Chichester: Wiley, 1987), pp161–74

11 Pringle, S, 'Sharing science', in Levinson, R, and Thomas, J (eds) *Science Today: Problem or Crisis?* (London: Routledge, 1997), pp206–23

12 Details of the events that have been organised to date in the European Weeks for Science and Technology are available from the European Commission, DG XII (Science, Research and Development), Brussels, Belgium.

13 For a well-informed review of this exhibit see Friedmann, A J, 'Exhibits and expectations', *Public Understanding of Science*, 4 (1995), pp305–13.

14 See 'Chemists sound warning on sponsorship of exhibitions', *Nature*, 380 (1996), p95 and 'Science in American Life', *Nature*, 380 (1996), p89.

15 See, for example, Weinberg, S, 'Sokal's hoax', *New York Review of Books* (8 August 1996), pp11–15 and references therein.

16 Piani, R, presentation to the *Here and Now* conference (London: Science Museum, 21 November 1996)

17 *Museum Collecting Policies in Modern Science and Technology* (London: Science Museum, 1988)

18 Keene, S, *Managing Conservation in Museums* (Oxford: Butterworth-Heinemann, 1996)

19 The term 'big science' was coined by Alvin Weinberg in the 1960s when he was director of the Oak Ridge National Laboratory, Tennessee, US. The term generally refers to hugely resourced, multi-national collaborations that often involve governments in their policy making. Examples are the Human Genome Project and projects in the space sciences, particle physics, oceanography and fusion research.

20 The case for the primacy of objects in museum displays has many eloquent advocates. See, for example, the comments on the new galleries in the American Museum of Natural History, New York, in Gould, S J, 'Evolution by walking', *Dinosaur in a Haystack* (London: Jonathan Cape, 1996), pp248–59.

21 Lowenthal, D, 'Material preservation and its alternatives', *Perspecta*, 25 (New York: Yale Architectural Annual, 1989)

22 For a general review, see 'Science versus anti-science?', *Scientific American*, 276 (1997), pp80–85.

23 Holton, G, *Science and Anti-Science* (Cambridge, MA: Harvard University Press, 1993), pp145–89

24 For a thoughtful discussion of how this might be done, see Bud, R, 'Science, meaning and myth in the museum', *Public Understanding of Science*, 4 (1995), pp1–17.

25 Contribution to the discussion 'What are we doing on-line?', *Harper's Magazine* (New York: August 1995), pp35–46

26 Turkle, S, *Life on the Screen: Identity in the Age of the Internet* (London: Weidenfeld and Nicholson, 1996)

27 Smith, A, *Software for the Self* (London: Faber and Faber, 1996), p107. This work contains a good deal of eclectic discussion about the continuing revolution in virtuality and interactivity.

28 For an interesting perspective on this issue, see Katz, J, 'Tom Paine and the Internet', in Spufford, F and Uglow, J (eds), *Cultural Babbage* (London: Faber and Faber, 1996), pp227–39.

29 Marshall, L, 'Code for a Grecian urn', *Wired (UK)* (September 1996), pp75–104

30 Greenfield, S, presentation to the *Here and Now* conference (London: Science Museum, 21 November 1996)

31 Many interesting perspectives on this question are given in 'When science becomes culture', in Schiele, B (ed), *World Survey of Scientific Culture* (Ottawa: University of Ottawa Press, 1994).

32 Shortland, M and Gregory, J, *Communicating Science: A Handbook* (New York: John Wiley and Sons, 1991)

33 For an account of the importance of demographic changes in the west over the next 20 years or so see McRae, H, *The World in 2020* (London: HarperCollins, 1994), pp97–119.

34 Rifkin, J, *The End of Work* (New York: G P Putnam, 1995), pp221–35

35 For example, see Blud, L, 'Sons and daughters: observations on the way families interact during a museum visit', *Museum Management and Curatorship*, 9 (1990), pp257–64.

36 An interesting discussion of young students' attitudes to science is given in Driver, R, Leach, J, Millar, R and Scott, P (eds), *Young People's Images of Science* (Buckingham: Open University Press, 1996).

37 Gago, J-M (ed), 'School science and the future of scientific culture in Europe', *Report to the European Commission* (1996)

38 See also the essays in Durant, J, and Gregory, J (eds), *Science and Culture in Europe* (London: Science Museum, 1993).

39 The key issues in this topic are accessibly reviewed in 'A survey of biotechnology and genetics', *The Economist*, 334 (25 February 1995).

40 Bauer, M (ed), *Resistance to New Technology* (Cambridge: Cambridge University Press, 1995), pp293–331

41 Lehrach, H, presentation to the *Here and Now* conference (London: Science Museum, 21 November 1996)

42 Birkerts, S, *The Gutenberg Elegies: The Fate of Reading in an Electronic Age* (London: Faber and Faber, 1994), pp117–64

43 White, S, Evans, P, Mihill, C and Tysoe, M, *Hitting the Headlines: A Practical Guide to the Media* (Leicester: British Psychological Association, 1993)

44 See, for example, Rheingold, H, *Virtual Reality* (London: Secker and Warburg, 1991). A more up-to-date review is given in 'Still a virtual reality', *The Economist*, 341 (12 October 1996).

45 Quoted in Smith, A, *Software for the Self* (London: Faber and Faber, 1996), p104.

46 Loomis, R J, 'Planning for the visitor: the challenge of visitor studies', in Bicknell, S and Farmelo, G (eds), *Museum Visitor Studies in the 90s* (London: Science Museum, 1993), pp13–23

Editors' note

Each of the papers in this collection is based on a presentation given at the *Here and Now* conference held at the Science Museum, London, from 21 to 23 November 1996. Copies of the transcripts of the discussions that followed each presentation at the conference are available from us on request.

It is a pleasure to thank all our colleagues at the Museum who assisted in the organisation of the conference, especially Rebecca Mileham, Rebecca Marshall and Claire Shaw, who all worked tirelessly to ensure the smooth running of the event.

Finally, we are indebted to Giskin Day who has copy-edited this volume with exceptional energy, thoroughness and patience.

Temporary contemporary:
some puzzles of science in action

Simon Schaffer

How can contemporary science be interpreted when the facts are still up for grabs? Can we reconcile history with morality in museum displays? This is an exploration of the slippery meaning of 'contemporary'.

In 1983 the American physicist Richard Feynman delivered a series of popular science lectures in Los Angeles. He started by complaining that 'people are always asking for the latest developments in the unification of this theory with that theory, and they don't give us a chance to tell them anything about one of the theories that we know pretty well. They always want to know things that we don't know.' When he published these talks on the interaction of light and electrons, he reported that, since the lectures, 'suspicious events observed in experiments make it appear possible that some other particle or phenomenon, new and unexpected (and therefore not mentioned in these lectures) may soon be discovered'. Then five months later he added that these (unknown) events 'appear to be a false alarm'. According to Feynman, in physics 'things change faster than in the book publishing business'.[1] Not just in physics, Mr Feynman, and not just in book publishing. Museums and science centres have the job of thinking about overwhelming visitor interest in 'hot topics' in science and technology. It is interesting, for example, that things seem to change faster in the sciences than in science museum displays. Feynman's typically pithy wit shows some of the puzzles this raises. It is imagined that the public wants to know about a frontier of ignorance constantly in flux, not a realm of secure knowledge well-entrenched in orthodoxy. The contemporary is torn between an understood, but tedious, past and an unknown, but enchanting, future.

Other enterprises in California, the home of shaky futures, show equally well how the contemporary culture of the sciences often challenges and subverts established expectations. The Exploratorium, founded in San Francisco in 1969 by the brilliant physics researcher and socialist educator Frank Oppenheimer, has now long inspired others to set up hands-on science centres worldwide. Most of these subsequent science centres, it has been acutely alleged,[2] give an image of the sciences entirely separated from social reality and consisting entirely of self-evident and lucid basic principles available to all. This was by no means Oppenheimer's original project. Historical propriety was there prudently reworked to make lessons accessible. Late twentieth-century artefacts, like cloud chambers from the National Aeronautics and Space Administration (NASA) and the spark chambers from Stanford, were ingeniously redeployed to give insight into the modern aftermath of Einstein's early twentieth-century relativity theory. And Oppenheimer's institution sometimes actively eschewed salient contemporary scientific hardware. Controversially for a northern Californian, he refused to put computers on show and was upset as electronics components became ever commoner in displays. Lasers and holography, endemic in most modern science museums and the stuff of contemporary science centres from Copenhagen to Canberra, suffered a less happy fate in San Francisco: fragile and pricey, they violated the Exploratorium culture. Its historian, Hilde Hein, notes that such electronics work obscures the processes of human labour: 'few outsiders can identify the workers' movements or participate in their intellectual activity'.[3] There were thus some rather good political and philosophical reasons to exclude the most obviously contemporary elements of late twentieth-century sciences from the world's most famous science centre. Recent work by Christine Brown and Gaynor Kavanagh shows, too, how crucial are gender and social relationships in the meanings and impacts of such hands-on exhibits.[4] In a recent analysis of the Science Museum's hands-on *Launch Pad* installations, John Stevenson reckons it is obvious that interactive physics exhibits are easier to set up than shows based on one of the principles of, say, politics. 'What would an interactive philosophical exhibit look like?' he asks.[5] Contemporary science centres often do pose philosophical puzzles with their exhibits, though, and any worthwhile show of contemporary sciences will evidently embody a choice of philosophical and political principles.

Here I focus on two technical, philosophical and political puzzles about the contemporary in the sciences. First, when scientific work is in process, as contemporary sciences necessarily are, then the very facts of the matter will still be openly in question. Before a vital debate closes, the relations of expertise and authority over the facts are also up for grabs. The lesson of recent science studies, and the problem for curators, is that authority and the facts are contemporaries of each other: one gets defined as the other does. So those keen to show up-to-date scientific work will find it hard to define sure relations of trust and authority. There is no reason to suppose that modern publics are incapable of creative responses to conflicted and debatable shows of science in action. Such shows will display open and personified arguments, often polemical, rarely securely grounded. Alan Morton has noted that the figure of Adam who stood on the facade of the Natural History Museum was not replaced when it fell during the Second World War.[6] I take Adam's fate as an emblem of the point that, faced with an open-ended pattern of a current science, even the most lapidary bit of a science museum might need to be reconstructed quite dramatically and quite soon. Second, I want to suggest that judgments of a piece of contemporary scientific work rely on long, drawn-out

stories about the past of that work. Never merely antiquarian, the investments and choices of the sciences' pasts dominate the ways they work now. Often, these histories will surprise, and will not be limited to the seemingly self-evident pathways the sciences pursue. Philip Doughty and Simon Knell, for example, have recently well explored the roller-coaster of museum geology, showing nicely how the Organisation of Petroleum Exporting Countries (OPEC) oil-price hike and subsequent fall retrospectively transformed the cultural memory of geology's past and its museological present.[7] Think of how Alfred Wegener's stories of continental drift[8] have themselves drifted from the margins to the centre of geology's self-representation during this century. Arthur Molella and Carlene Stevens have graphically described how, in 1989, chemists congratulated the Smithsonian Institution curators for quickly collecting material on cold fusion; a few months later, as Pons and Fleischman's stories fell from grace, the curators were roundly damned as subversives for even mentioning, let alone displaying, this very set-up.[9] The contemporary has a dynamic history strongly linked with rival moralities.

The passion of such exchanges shows that we must not restrict ourselves to asking how the contemporary sciences can be made lucid to the wider culture, as though understanding that culture is easy and as though the sciences alone possess a dynamically fluid quality. It is often complained that a problem of showing contemporary sciences is that it is hard to see what makes them visitors' contemporaries, what features of everyday life and labour are to be matched up with this or that gleaming new device. We surely also thus have to ask which aspects of any culture are relevantly contemporary to the sciences of the time and ask, too, about the mediations between these bits of culture. For example, the audiovisual programme which accompanies the *Science in the 18th Century* gallery at the Science Museum, London, carefully shows cultural, economic and political features of immense importance for the eighteenth-century artefacts on show in the adjoining rooms. Despite well-publicised prejudices which too often urge the contrary, skilful historical and curatorial judgment can and must make out the contorted entanglements of the social in the scientific. I conclude that the openness and the historicism of contemporaneity both pose challenges for the showing of sciences. I propose in conclusion that creative strategies of juxtaposition are important for improving the presentation and production of contemporary sciences.

News of meteorite ALH84001 and its implications for the existence of life on Mars neatly show how these twin processes of contemporary sciences work. After 4 billion years beneath the Martian surface, 16 million years in interplanetary space, 13 millennia in the Antarctic, eight years of misclassified storage as a piece of diogenite in a fridge in the Johnson Space Center, Houston and then two years' hard collaborative work by geologists, chemists, spectroscopists and exobiologists, this rock reached the public gaze during a press conference on 7 August 1996. The performers there pointed out the carbonate rosettes in the meteorite, surrounded with polycyclic hydrocarbons and, where the carbonates had dissolved, magnetite crystals of characteristic shape. They added that there was electron microscopic evidence of nanofossils. Four elements of a story, then, each separately defeasible, but in combination suggesting a tale of early Martian life. It's characteristic of such claims that they possess the character of a narrative and never simply depend on the work of pointing to a brute object for the signs of truth. Ten days after the story broke, the curators of the Science Museum's *Exploration of Space* gallery installed a small show about the meteorite featuring newspaper headlines alongside more sober quizzes about the rock's meanings. If curators have typically searched for the objects to include in the object lessons of their displays, then here's a fine example of how the objects must be made elements of continuous contemporary stories. This complex story of meteorite ALH84001 soon became entangled with other tales. In any exhibit about life on Mars, we need at once to think about how a choice between these contemporary tales could so easily destabilise one version of the facts and the authorities, yet privilege others.

Simple journalistic enthusiasm about the possibility of life on Mars was to be expected. Political exigencies and funding crises at NASA help explain the verbiage of President Bill Clinton: 'Rock 84001 speaks of the possibility of life. If this discovery is confirmed, it will surely be one of the most stunning insights into our universe that science has ever uncovered. Its implications are as far-reaching and awe-inspiring as can be imagined. Even as it promises answers to some of our oldest questions, it poses still others even more fundamental. We will continue to listen closely to what it has to say as we continue the search for answers and for knowledge that is as old as humanity itself but essential to our people's future.'[10] Predictable, too, were the soundbites of well-known science publicists. 'Every American and every human should take pride in this accomplishment of NASA.'[11] Thus Carl Sagan, Cornell exobiologist and founder-president of the Planetary Society, who also opined that 'if the results are verified, it is a turning point in human history'. The prolific Adelaide natural philosopher Paul Davies, now writing a book about life on Mars, announced that such a discovery would be more significant than those of Copernicus, Darwin or Einstein: 'the most amazing of all time'. Davies is reported as claiming, 'I told you so,' referring to his 1993 book *Are We Alone?*, and guessed enthusiastically that it was possible that 'we would all be descended from Martians'.[12] This is a credible fantasy of which H G Wells, the ambiguous theorist of alien life, would have been proud.

So here was an event of obvious public appeal and thus, presumably, a candidate for careful museological display. But who to trust? One rule which well applies to scientific confidence is that it tends to increase with social distance from the claim in question. Shows for a wider public often express more security than debates within a specialist community. The NASA team's leader told the August press conference that 'this is a controversial story and there will be lots of disagreement'. He was not wrong. William Schopf, a palaeobiologist from the University of California, Los Angeles, publicly doubted that organic remains in the meteorite could come from biological processes; noted that colonies of classic cell reproduction had not thus far been located; and pointed out that the researchers' 'microfossils' were implausibly small.[13] Evidence of such 'nanobacteria' on Earth is notably controversial, sometimes attributed to artefacts made in the very process of preparing the samples for electron microscopy. Any show which included displays on the artefacts that infest microscopic technique would draw attention towards this worry, and thus away from the easy identification of the nanofossils. Jim Papike and Chip Shearer, meteorite experts at the University of New Mexico, had tested the same meteorite's sulphur components for biomarkers but found no evidence of biological activity.[14] Is the shape of its much-vaunted magnetite crystals a sign of life? 'Shape is one of the worst things you can use in geology to define things' says the Washington State geologist Everett Shock. Datings from Chicago's Field Museum and from the Scripps Oceanographic Institute give an age for the tell-tale carbonates not of 3.6 but 1.3 billion years, a period long after water ceased flowing across the Martian surface. Wisconsin and Oregon biologists reckon carbonate rosettes could well be found in non-biological settings, while Ralph Harvey, the Case Western Reserve scientist who helped process the meteorite in the first place, now says the carbonates were produced by very high carbon dioxide concentrations after a high-temperature asteroid crash on Mars. Derek Sears, editor of *Meteoritics*, reckons that the NASA team's precautions against contamination were 'flaky and simplistic'. Many reckon that no firm judgment on such Martian life is possible without further landings on the Red

Planet—back to the cruder motivations of boosterism in advance of successive NASA Mars shots.[15] In its last issue of 1996, *New Scientist* ran the headline 'Death knell for Martian life'.

Interviewed in a superb BBC *Horizon* documentary on the meteorite controversy, co-authored by Danielle Peck and Max Whitley, Schopf declared that 'science is not in the belief business', and that for such high stakes 'irrefutable evidence' is indispensable.[16] Yet here, belief and its refutation are clearly the very stuff of the living process of contemporary sciences. Richard Zare, a Stanford chemist in the NASA team, wonders 'who is to say we are not all Martians?'[17] Who indeed? So here is a basic trouble of the contemporaneity of the sciences. Authority and the facts are contemporary and get fixed together. This does not mean that anything goes. On the contrary, in the messy real-life world of contemporary sciences, prior judgments of plausibility count. This is why, for example, I have been careful to mention the institutional affiliation of each protagonist in the Mars debate. When Paul Davies set out in 1993 to discriminate between his own belief in invisible dark cosmic matter and his denial of oft-visible UFOs, for example, he said the difference between scientists and cranks was that scientists build on existing sciences—and, we might add, this means building on existing networks of belief in specific persons as much as in special things.[18] The implication for science museums keen to show contemporary sciences is that, willy-nilly, the past of the sciences, and the track record of scientists and their institutions, matter to any account of the plausibilities of the sciences now.

The problem of biological remains in meteorites is not new, but has a long history that repeatedly shows the contemporary establishment of authority and fact. Meteorites, just like any other scientific object, get their meanings within specific and very long-term stories. Such objects, it's been said, have biographies. And because narrative gives objects meanings, no such objects can well be understood in simple-mindedly contemporary terms. Enlightenment orthodoxy once confidently rejected the possibility of stones falling from the sky. Antoine Lavoisier and other eminent natural philosophers in 1772 damned the idea as a popular fallacy to be ranked with other peasant superstitions 'which can only excite the pity not only of natural philosophers but of all reasonable people'.[19] Reports of phenomena unexplained by the experts received from the inexpert should, it was supposed, be rejected. Such phenomena had to be produced and reproduced within authoritative settings before they became credible. Their extraterrestrial origin accepted, meteorites then became the target of expert chemical and cosmological work. George Wilson, first director of the Scottish Industrial Museum in the 1850s, was one among many who reckoned that the chemical composition of meteorites would provide information about extraterrestrial life. The pre-eminent Scottish natural philosopher Lord Kelvin even judged that life must origi-nate from their impact on Earth, though his 'star germ' theory drew ribald criticism from Charles Darwin's bulldog Thomas Henry Huxley.[20] In 1881 Otto Hahn, a German geologist-turned-lawyer, won enormous pub-licity from the star germ theorists for his pic-tures of microfossils in ground-up slices of meteoritic chondrites. The scandal surround-ing his stories and images was not forgotten— Hahn's work was recalled in 1962 by the Chi-cago scientists Edward Anders and Frank Fitch when they, in turn, set out to demolish claims that paraffinoid hydrocarbons, and even fossil algae, had been found by Bartholomew Nagy's group of petroleum scientists in carbonaceous meteorites from France. The eminent crystallographer and socialist polemicist Desmond Bernal, long a fan of meteoritic theories of life's origins, influentially backed Nagy's findings. But 'the decision whether a certain form is of biological or inorganic origin', the Chicago experts riposted, 'is quite subjective'.[21] And the fate of Nagy's claims has often been mentioned dur-ing this year's meteorite furore. There was no moment in the 1960s and 1970s when an easy or objective consensus was established on the problem. Nagy remained fairly sure that a biological interpretation of such meteorite components was viable; his erstwhile patron the Nobel laureate, Harold Urey, reckoned contamination problems might be explained away as results of meteorites' lunar origin; and new findings in Australia and Antarctica, processed at NASA's immensely prestigious Ames Research Center after 1970, seemed to hint that such biologically crucial substances as amino acids could well be made in deep space.[22] At each stage in history, contemporary scientific culture rewrites its various pasts' meanings, then uses those versions to make some claim stick in the present.

So to evaluate the plausibilities of our exactly contemporary attitudes to meteorite ALH84001—an unremarkable if long-distance rock or the key to human destiny—needs a choice between a number of different stories. These include long histories of elite attitudes to meteorite samples, narratives of space exploration including the 1976 Viking mis-sions which seemed so decisively to rule against a habitable Martian environment, stories which look forward to the Martian Global Surveyor of 1997 (Clinton was careful to mark that its scheduled arrival is the Fourth of July) and stories which look backward beyond the location of the meteorite in early 1984. On even the most restricted version of this rock's sense, we may wish to draw atten-tion to the funding implications for NASA and the Russians, to the role of geologists in amassing meteorite samples, to the possibility of extraterrestrial life, or to the high-tech skills of spectroscopy and scanning electron microscopy. Each story and each setting sug-gests rather different attitudes to the rock in question and the means to be used to display its meaning, and each also links up with different relations of authority and trust, ex-pertise and belief.

Because any evaluation of exactly contem-porary sciences in process requires the choice of a history in which to place them, museums must willy-nilly make interventions in the mediations of culture and the sciences. In her wide-ranging survey of science and technology museums, Stella Butler has rightly insisted that the inseparability of contemporary culture and its sciences must be a central theme of any viable display.[23] She also notices the apparent mismatch between this laudable aim and the demands of museum visitors, who 'expect to experience spectacular experiments, to see apparatus associated with the most significant scientific discoveries and to marvel at the engineering techniques of modern industrial society'. She reckons that 'the fascination of the general public with the past shows no signs of abating'. At the very least, critiques of the heritage industry will carry over to scepticism of shows of past sciences, for museums have not always found it easy to pick out compelling and challenging accounts of the national past. The problem is all the more acute in shows of contemporary sciences and their material culture. Ruesselsheim's city museum boldly shows its local firm Opel's past links with National Socialism: such frank commitments are less apparent elsewhere in the museum community.

Obviously there is a slightly self-serving aspect to my argument: in order to show contemporary sciences we need a historical understanding of the meaning of 'contemporary' alongside a grip on the paths of the sciences. I ply my own trade in a department of history and philosophy of science built round an earlier, more distinguished, museum: the Whipple Museum of the History of Science. I cannot, therefore, argue against the significance of the museum as a site of contemporary knowledge production. Indeed, for a historian this is no news. A wealth of studies exists on curiosity and collection, on natural history and the Wunderkammer, on civil conversation and patronage, which shows us that a range of early modern programmes, including chemistry, astrology, natural history, antiquarianism and connoisseurship, were produced in-and-as the work of the Renaissance and Baroque museum. The very

term 'contemporary' entered our language at the same time as the foundation of England's first public museums in the mid seventeenth century.

The fascination of coins and medals, with its accompanying culture of curiosity, was assailed and displaced by encyclopaedist promoters of philosophical natural history in the first half of the eighteenth century. During the nineteenth century, experimentalists stressing control over nature's capacities played down the significance of the museum as a site of scientific work and, as John Pickstone has shown, 'those who advocated laboratories tended to see museums as second-rate—useful for the public but not the temples of science'.[24] Part of the success of the Exploratorium was precisely that its installations mixed laboratory and museological elements, and treated display spaces as places where new knowledge and technique could be produced. The theme has a wider political meaning: at the Zurich Heureka show in 1991, its architect Georg Mueller made sure to produce spaces where active public dialogues on gene technology, solar power, ecology and health could flourish.[25] In such places the meaning of 'contemporary' was reinterpreted in refreshingly new directions: the public were seen as genuine contemporaries of sciences in action, not merely consumers of finished techniques.

There is therefore a powerful history of the contemporary. A decade ago, Ray Batchelor reflected on tensions at the Science Museum between the sophisticated nostalgia of the modish room-set and the absolute need historically to document the many resonances of even the most mundane object. Batchelor neatly showed how a commonplace Edwardian kettle could be linked with its manifold contemporaries from tariff reformers to kitchen designers—the effect is at once surreal and exciting.[26] To do justice to these resonances, we need accounts of changing notions of the contemporary. Here are two examples. For Enlightenment Europeans, the contemporary was often located rather far away from home.

Jonathan Swift and Voltaire both examined their contemporary forms of natural knowledge by imagining distant lands elsewhere on Earth or on other planets. We could use this point to reorganise the layout of shows of eighteenth-century sciences as part of the movement of voyaging and performing. As the Australian historian Greg Dening shows, expert Polynesian navigators were the more ancient contemporaries of the Enlightenment Europeans who entered the South Seas bearing longitude clocks and astronomical instruments, who then exhibited these Polynesians in the theatres and museums of London and Paris.[27] So a show now of the clocks of John Harrison and Larcum Kendall at the National Maritime Museum, Greenwich, might well juxtapose them with devices which our current division of cultural labour preserves in the South Pacific collections at Greenwich, the British Museum, the new national museums of Pacific nations, or in the theatre galleries of the Victoria and Albert Museum. This would help disturb some taken-for-granted prejudices about the sciences of time and space then and now. Or, for example, think about the way in which H G Wells's works are effortlessly used in stories about Martian life. In 1898, when *War of the Worlds* appeared, European contemporaries saw a technologically militant bourgeoisie confronting the still active past of a ruling class profoundly aristocratic and traditionalist, and the potent future of working-class insurrection and imperial warfare. Wells's Martian story is simultaneously a tale of alien life and a tale of imperial conquest, with the Martians playing colonists to the Earthlings' natives.[28] Thus a display which recalls that Wells's Martians were contemporaries of those of the visionary astronomer Percival Lowell would need to use resources from the Imperial War Museum alongside those of South Kensington so as to juxtapose empire and the sciences. These kinds of juxtaposition are the crucial elements in reworking the meaning of contemporaneity hand-in-hand with the meaning of the sciences.

The presiding genius of South Kensington,

Wells, provides me with my punchline. The image of museums as time machines is now a commonplace. This does not merely mean that such institutions are the prisoners of their past: they obviously offer resources for changing time management in the present and reworking conceptions of the future.[29] In the midst of his journey to the world of 802,701 AD, when class differences had ineluctably evolved into the cannibalism of Morlocks and Eloi, Wells's Time Traveller briefly finds himself in a vast exhibition. Stretching over the Surrey hills of this late Victorian future is a Palace of Green Porcelain, 'some latterday South Kensington'. It contains sections of palaeontology, geology, natural history, a gallery of machines, technical chemistry, a library, an armoury and model mines. 'Everything had long since passed out of recognition. A few shrivelled and blackened vestiges . . . that was all!' It is the scientific metropolis of Wells's student days, its categories precisely preserved 800 millennia after its literary appearance in 1895, but massively enlarged, and then eventually allowed to fall completely into ruin: the curator's nightmare. The functions of this museum are complex. The Traveller tries to find himself there, he searches for weapons to use against the barbarous Morlocks who have already seized his own vehicle. Like any moderately vain scientist, he searches the library for his 17 papers on physical optics, once published in the *Philosophical Transactions of the Royal Society* but now reduced to a 'sombre wilderness of rotting paper'. Oddly, it does not occur to him to seek his own time machine in the engineering gallery where, logically, it should surely have found a pre-eminent place in any museum of the future. He confesses his 'certain weakness for mechanism', viewing these future engines as 'puzzles, and I could make only the vaguest guesses at what they were for'.[30] Wells makes the Palace into a brilliant and perverse symbol of what time machines can and cannot do. The Palace, after all, is precisely a huge machine for preserving the present as it moves into the future, while the time machine is a small museum of then contemporary science and technology. Young Wells was once a pessimist about the future of his culture and the meanings of its sciences, but through his surrealism he well showed how science museums helped make contemporary meanings. Surrealism is precisely the creative and ingenious juxtaposition of apparently unrelated, but oddly contemporary, objects of modern technologies and cultures.[31] That is a good recipe for science shows. My optimistic if equally surreal aim has here rather been to contemplate the ways in which such museums and machines can creatively interact to produce better and more honest stories of our own contemporary world.

Notes and references

1 Feynman, R P, *QED—The Strange Theory of Light and Matter* (London: Penguin, 1990), pp3, 152

2 For a critique of science centres see Durant, J, 'Introduction', in Durant, J (ed), *Museums and the Public Understanding of Science* (London: Science Museum, 1992), p10.

3 Hein, H, *The Exploratorium: The Museum as Laboratory* (Washington: Smithsonian Institution Press, 1990), pp32, 55–57

4 Brown, C, 'Making the most of family visits: some observations of parents with children in a museum science centre', *Museum Management and Curatorship*, 14 (1995), pp65–71; Kavanagh, G, 'Dreams and nightmares: science museum provision in Britain', in Durant, J (ed), *Museums and the Public Understanding of Science*, pp81–88

5 Stevenson, J, 'Getting to grips', *Museums Journal*, 94 (May 1994), pp30–32

6 Morton, A, 'Tomorrow's yesterdays: science museums and the future', in Lumley, R (ed), *The Museum Time Machine* (London: Routledge, 1988), pp128–43; Shapin, S, 'Why the public ought to understand science-in-the-making', *Public Understanding of Science*, 1 (1992), pp27–30

7 Doughty, P S, 'Museums and geology', in Pearce, S (ed), *Exploring Science in Museums* (London: Athlone, 1996), pp5–28; Knell, S, 'The roller-coaster of museum geology', in Pearce, S (ed), *Exploring Science in Museums*, pp29–56

8 See Stewart, J A, *Drifting Continents and Colliding Paradigms* (Bloomington: Indiana University Press, 1990).

9 Molella, A and Stevens, C, 'Science and its stakeholders: the making of "Science in American Life" ', in Pearce, S (ed), *Exploring Science in Museums*, pp95–106

10 Transcript of Presidential press conference, US Newswire Service (7 August 1996)

11 Friedman, L D, 'Sagan calls evidence on life on Mars possible turning point in human history', *Planetary Society News Release* (7 August 1996)

12 Dayton, L, 'Life on Mars? Scientists now believe it's possible', *Sydney Morning Herald* (8 August 1996), p1; Davies, P, 'Why we may once have been Martians', *Sydney Morning Herald* (8 August 1996)

13 Carreau, M, 'Mars rock: a migration for the ages', *Houston Chronicle* (8 August 1996)

14 Fleck, J, 'Similar UNM meteorite study found no evidence of Mars life', *Albuquerque Journal* (8 August 1996)

15 Beatty, J K, 'Life from ancient Mars?', *Sky and Telescope* (October 1996); Gibbs, W W and Powell, C S, 'Bugs in the data?', *Scientific American*, 275 (1996), pp12–13; Fleck, J, 'Similar UNM meteorite study found no evidence of Mars life'

16 Peck, D and Whitley, M, 'Aliens from Mars', *Horizon* (London: BBC, 11 November 1996)

17 'Was there once life on Mars?', Space Views: Breaking News, 9 August 1996 (World Wide Web page by National Academy of Science)

18 Davies, P, 'A window into science', *Natural History*, 102 (1993), pp68–71

19 For objects' biographies see Appadurai, A (ed), *The Social Life of Things: Commodities in Cultural Perspective* (Cambridge: Cambridge University Press, 1986) and Briggs, A, *Victorian Things* (London: Penguin, 1990). For meteorites see Westrum, R, 'Science and social intelligence about anomalies: the case of meteorites', *Social Studies of Science*, 78 (1978), p470.

20 Crowe, M J, *The Extraterrestrial Life Debate 1750–1900* (Cambridge: Cambridge University Press, 1986), pp328, 400–6

21 Dick S J, *The Biological Universe: The Twentieth Century Extraterrestrial Life Debate and the Limits of Science* (Cambridge: Cambridge University Press, 1996), pp371–72

22 Crowe, M J, p406; Dick, S J, pp368–72

23 Butler, S, *Science and Technology Museums* (London: Leicester University Press, 1992), pp120, 131–32, 57

24 Pickstone, J V, 'Museological science? the place of the analytical-comparative in nineteenth-century science, technology and medicine', *History of Science*, 32 (1994), p132; Arnold, K, 'Presenting science as product or as process: museums and the making of science', in Pearce, S (ed), *Exploring Science in Museums*, pp57–78

25 Philips, D, 'Heureka! it's hands on', *Museums Journal*, 94 (May 1994), pp28–29

26 Batchelor, R, 'Not looking at kettles', *Museums Professionals Group News*, 23 (1986), pp1–3

27 Dening, G, *Performances* (Melbourne: Melbourne University Press, 1996), pp207–24; Dening, G, 'The theatricality of observing and being observed: eighteenth-century Europe "discovers" the ? century "Pacific" ', in Schwartz, S B (ed), *Implicit Understandings: Observing, Reporting and Reflecting on the Encounters between Europeans and Other Peoples in the Early Modern Era* (Cambridge: Cambridge University Press, 1994), pp451–83. For Enlightenment notions of the contemporary see Outram, D, 'On being a contemporary', in Clark, W, Golinski, J and Schaffer, S (eds), *Sciences in Enlightened Europe* (Chicago: Chicago University Press, 1997).

28 Haynes, R D, *From Faust to Strangelove: Representations of the Scientist in Western Literature* (Baltimore: Johns Hopkins University Press, 1994), pp165–66

29 Silverstone, R, 'The medium is the museum: on objects and logics in time and space', in Durant, J (ed), *Museums and the Public Understanding of Science*, pp34–42

30 Geduld, H M (ed), *The Definitive Time Machine* (Bloomington: Indiana University Press, 1987), pp72–76; Schaffer, S, 'The Time Machine', in Bennett, J, Brain, R and Schaffer, S *et al.* (eds), *1900: The New Age* (Cambridge: Whipple Museum, 1994), pp13–16

31 For a good account of the surrealist juxtaposition of objects of modern science and technology, see Cardinal, R and Short, R S, *Surrealism: Permanent Revelation* (London: Studio Vista, 1970), pp13, 33.

Contemporary science in science museums—a must

Wolf Peter Fehlhammer

Science museums have no choice but to present contemporary science. How is the Deutsches Museum tackling the problems?

The challenge of presenting contemporary science is now being taken up by several European science museums as they put together plans to expand their public areas. The Science Museum, London, is increasing its area by 30 per cent, Museu de la Ciencia, Barcelona, will triple in area, and the Deutsches Museum, Munich, is also planning a 20–25 per cent increase. Such activity goes some way to refuting the claims of some that museums are not a growth market, but the challenge still remains: what should be done with the new space once it has been secured? Should it be used to present contemporary science? And if so, how can this be achieved? I will attempt to answer these questions by considering, in the context of the Deutsches Museum, what the challenges are and how they can be met.

Plans to expand the Deutsches Museum

On 23 May 1996, the Bavarian Prime Minister delivered a speech in which he promised the Deutsches Museum 19 million marks (£7.5 million) for a 'transport museum' to be built on the site of an old fairground. The plan is to move the existing train and car exhibitions from the Deutsches Museum's Isar Island site (figure 1) to the fairground in order to create an ultramodern museum of mobility, particularly of information, but also of people, goods, materials and energy. The Prime Minister's speech was one of the greatest moments in the history of the Deutsches Museum, and his proposal underlines the importance of making way for 'new science' in museums. As he pointed out: 'For our children and grandchildren, the charms and marvels of technology go far beyond the mining and railway engines of today.' This most recent support for presenting contemporary science and technology follows on from a strong tradition of such activity in Munich, as exemplified by popular lectures.

Popular lectures

The tradition of popular lectures can be traced back to the seventeenth-century public lectures which were accompanied by demonstrations

Figure 1. The entrance to the Isar Island site of the Deutsches Museum, Munich

of physics and chemistry experiments. While the Academy of Science in Munich cultivated the idea,[1] it was Justus von Liebig (1803–1873) who really made these shows popular during his first year in Munich in 1853. An unplanned explosion during one of the lectures which wounded and nearly killed members of the Bavarian Royal Court, ensured that Liebig's fame soon spread beyond the city. Luckily, the Royal Family did not hold the incident against him; indeed, he became even more sought after at court, and chemistry lectures in Munich gained a popularity that they still retain today.[2]

Liebig's enthusiasm for popularising science owed much to the influence of Alexander von Humboldt's lectures, particularly the 'Kosmos' lecture series in Berlin. It was here that von Humboldt's vision for national science education led to the founding of Urania, a 'palace' for popularising science with its own astronomical observatory, a specially equipped scientific theatre, lecture rooms, laboratories and museum collections. A series of these centres was subsequently established throughout Europe and this inspired Oskar von Miller to found the Deutsches Museum in 1903. The venerable Conservatoire des Arts et Métiers in Paris and the Science Museum, London, were particularly influential in his thinking.[3]

Since its inception, the Deutsches Museum has organised popular lectures modelled on Liebig. At first these were sporadic, but given by famous scientists such as Jacobus van't Hoff and Carl von Linde. They then became more frequent, with Professor Hermann Auer delivering a very popular biweekly series on physics.

After a break of about 15 years, the Deutsches Museum resumed its popular lecture programme during the winter of 1993/94 with two talks by the Nobel laureate Roald Hoffmann, a chemist and gifted writer. These were followed by a third event, 'Chemistry Imagined', when Hoffmann's poems were combined with collages by the artist Vivian Torrence to produce an exhibition—unique for a science museum—that brought together the polarised cultures of art and science.

The Museum invited further Nobel laureates to open the summer Science Week in 1994, and a successful winter lecture series followed on from this. The popularity of these lectures, which were often accompanied by demonstrations of experiments, appeared to lie more with the physical presence of the scientists and the opportunity to ask them questions than with the actual topic of the talk. Recently, two talks were given on the contraceptive pill and the 'morning-after pill', RU486, by Carl Djerassi and Etienne-Emile Baulieu, respectively. It was the first time that we dared to tackle such controversial issues. Baulieu's fears that the largely Catholic Bavaria would protest violently proved unfounded, and mercifully his bodyguard was unnecessary.

Judging from the audience's enthusiastic response and the regular overcrowding of the lecture hall we should have been quite happy. However, looking at the numbers brings us down to Earth quickly: we reach just over 5000 people per year this way, which is less than 0.5 per cent of the total visitors to the Deutsches Museum. It follows that popular lectures cannot serve as the Museum's main means of disseminating science to the public, even if we were to publish them. The daily demonstrations taking place in the chemistry gallery since 1969, on the other hand, are attended by an audience of some 25,000 per year. By proceeding from weekly or biweekly popular lectures, to lectures with experiments, to daily experimental lectures in the exhibition space, we seem to be on the right track. In order to fulfil our educational mission efficiently, we are now aiming at hourly lectures on the galleries.

Reasons for presenting contemporary science and technology in science museums

Initially, most if not all science museums and academies were conceived of as institutions with collections of contemporary artefacts. We should therefore strive to keep up to date, not least out of respect for our founders.

The original purpose of the Deutsches Museum has been much debated, although evidence suggests it was founded with the intention of interpreting the present, or even the future. Oskar von Miller's ambition was for it to proclaim the glorious rise of the engineering profession, and to demonstrate that there was a wealth of novel ideas and visions in German science and technology. With regard to von Miller's desire to emphasise Bavaria's promotion of trade and industry, as outlined in a 1903 circular,[4] Radkau pointed out how the early Deutsches Museum, built in rural and conservative Bavaria, was even more extraordinary than von Miller had envisaged in that it highlighted Bavaria's wealth of technological culture.[5] What sounded optimistic then has now long been taken for granted. After the Second World War, formerly rural and conservative Bavaria became Germany's most high-tech state, with Munich a high-tech metropolis; a contradiction which, since the time of King Ludwig I, has tended to become entrenched in government policy.[6]

Attitudes to presenting contemporary science in the Deutsches Museum can be gauged by comparing the educational mission of von Miller's 1903 proclamation with that of the statutes.[7] While von Miller was prepared to take socio-political influences into account, he insisted that the Museum should present up-to-date and contemporary science, an aim that Prince Ludwig himself strongly supported. In the statutes, however, the deletion of every statement or word pertaining to anything contemporary is striking. Perhaps there was a sudden dominance of historians on the Board.

A second reason for displaying contemporary science in museums is the inherently forward-looking nature of the scientific discipline. This had already been recognised in the proposals for the new chemistry gallery in the 1970s which stated that it should contain as much as 40 per cent on current topics, with 30 per cent on history, and 30 per cent on 'edutainment'. And a refusal to 'go

contemporary' is more contradictory to the nature of science than to the nature of literature.[8]

Extending the educational mission

A second and psychologically more powerful approach to understanding the importance of contemporary science in museums is to ask what would be lost if it were excluded.

First, certain groups of visitors would stop coming; for example, the 16-year-old computer enthusiast, who is already frustrated when galleries do not include the latest equipment. Second, arrangements between museums and the formal education system would break down. Schools have become aware of the importance of incorporating the informal education system, such as visits to science museums, into their curricula.[9] Schools currently provide some 30 per cent of our visitors. Third, science museums would no longer be involved in the important process of contributing to lifelong learning and promoting Europe as a 'learning society'.[10] This would contradict the aims of both the Deutsches Museum and the European Collaborative for Science, Industry and Technology Exhibitions (ECSITE). In addition, science museums would not be able to have a role in providing individuals with an opportunity to come to terms with the science and technology of their own society. Negative attitudes to science range from simple ignorance through non-acceptance to militant hostility. Evidence of this is emerging not only in Germany,[11] but also in other European countries, the US and Japan. Ultimately, the economy, employment and standards of living are threatened.

The reasons for these negative reactions are complex, but they are almost all traceable to contemporary rather than historical science. Science museums have an extremely valuable role to play in presenting up-to-date science in an objective way, so that people may make up their own minds. According to Helmut Schmidt, the former German Chancellor, many of the public's anxieties about science

Table 1. Vision 2003

Philosophy/mission statement

- topicality in all activities
- primacy of exhibits
- collecting not an end in itself
- extended educational mission
- deal with controversial subjects
- living forum
- foundation of satellite museums

Exhibitions

- plurithematic
- integral, dynamic
- popular, entertaining, exciting
- 'transient clusters' of performers

Collections

- defining a collections policy
- installing a collections management system

Research

- scientific excellence
- intensification of object-based research
- integration in Museum's life
- editing of a first-rate book series (in collaboration with other museums)

Marketing

- visitor-orientation in all activities

Funding/organisation

- partial privatisation (of secondary activities)
- new sources of income
- creating a foundation trust 'Deutsches Museum'
- external placing of orders
- installation of project management in all central areas

and technology may be assuaged by 'making clear which anxieties are justifiable, and which are not . . . this is one of the most important tasks for science, or . . . scientists'.[12] The Deutsches Museum is addressing these negative attitudes by working with consultants to draw up a new mission statement, *Vision 2003*, which outlines the key points of a series of half-year projects (table 1). In the context of contemporary science, the move towards entries such as 'extended educational mission', and the intention to present 'controversial themes' should be noted.

These new strategies for enabling people to come to terms with developments in science and technology are being attempted in the new chemistry gallery. This is particularly apt, as chemistry has now almost replaced nuclear energy as the 'bogeyman' in Germany. It also generates hostile reactions in other parts of Europe. In response to this hostility, ECSITE has set up the 'Chemistry in European Museums' Project which formulated key 'messages' at a 1995 Barcelona meeting that will be used to structure the new chemistry gallery at the Deutsches Museum (figure 2). These signal a major departure from the text-book approach, and will involve the public in a more direct and informal way. Following in the tradition of public lectures, the gallery will host several live demonstrations every hour.

Challenging the sceptics

One approach to answering the question of how science museums can present contemporary science is to counter the criticisms from those who believe that museums are not the right forum to address these issues. Some argue that there is a danger in presenting scientific discoveries before they have been confirmed beyond all doubt. But the risk of error and misinterpretation is an inherent and important part of scientific discovery, and a museum is a good medium to illustrate this. An exhibition on scientific and technological failures or inaccurate predictions such as polywater, cold fusion and 'vertical take-off'

Not even chemists are perfect!

Figure 2. One of the 'global messages' formulated by ECSITE that will be used to structure the new chemistry galleries at the Deutsches Museum

might make science and scientists more accessible to the general public. Other criticisms include that modern science is too complex for museums to be able to cope with it. There are also concerns that museums might be misused as vehicles for promoting industry without being sufficiently objective.

A more fundamental approach to the problem is to ask whether science museums and science centres do in fact succeed in promoting at least some understanding of science. Extensive research using opinion polls and visitor behaviour in museums indicates a similarity between the activities of shopping and visiting a museum.[13] This view is supported by Miles, who suggests that it is 'an erroneous belief that science can be communicated to a lay public simply by delivering the information in an efficient way'.[14] But he does go on to describe 'different levels of involvement', which recognise that the visitor does not necessarily need to understand a scientific phenomenon in order to be familiar with it. Achieving that familiarity is an important part of the understanding process. Other factors have a role too, such as whether the museum makes the visitor feel comfortable. Miles does offer some hope in his conclusion by conceding that 'museums do present opportunities for awakening people's interest in a

subject', and that 'things can be enjoyable challenges' in museums, especially when they relate to the visitor's everyday life.

Extreme ways of presenting contemporary science

A method of exhibiting the cutting edge of science and technology is to demonstrate real research and production in the galleries themselves. For example, the *Emballages* (Packaging) exhibition at La Cité des Sciences et de l'Industrie, Paris, presented a real packing plant using a mixture of multimedia and working machinery. Another possibility is to involve staff directly in giving lectures or undertaking research on subjects of current interest. Museum staff who are involved at universities can benefit from belonging to the scientific community through access to advice and sophisticated equipment, and through coming into contact with a useful source of speakers and motivated young students (always a potential supply of suitable explainers).

The idea that university professors can be senior museum curators is a familiar one. Three members of the Deutsches Museum staff hold the title of 'professor'. We are now trying to bring some coherence to the research programme of the Museum with *Vision 2003*. This states that all research proposals must be submitted to a committee that will decide whether to grant a sabbatical under the supervision of the Museum's research institute. This will ensure that research is of the highest quality. It also guarantees that the Museum will be recognised as research-active, which is important because research contributes 30 per cent of the Museum's income. Further recognition will be achieved with the publication of a book series, starting in 1998, provisionally entitled *Technology, Museum Collections and Material Culture* in collaboration with the Science Museum in London and the National Museum of American History in Washington.

An obvious way to present contemporary science is to work in the opposite direction: bring university scientists into the museum. The Deutsches Museum initiated such a project in collaboration with the Ministry of Science in 1995. Scientists working for the Programme for Advancing Research and Technology Transfer in Bavaria were invited to promote their work in a forum hosted by the Museum, called *Aktuelle Wissenschaft*. Scientists ranging from postgraduates to research associates to professors produced exhibitions in collaboration with the Museum. The event proved once again that, as far as scientific research is concerned, it is the producer (scientist) rather than the product (the research) that really engages and inspires audiences. As Wengenroth said, the public's lost confidence in science and technology must be replaced by one in scientists and technologists. This will only work, however, if the researchers see themselves not as gods but as parish priests.[15]

Scientists available for visitors 'to touch' certainly takes the concept of interactivity to its extreme, and this is clearly a successful way of making science more approachable. Bringing scientists down from their ivory towers and presenting them as fellow human beings is an approach also employed in the *Science for Life* exhibition at the Wellcome Centre for the History of Medicine, London, where visitors are asked to guess which young people will become scientists by looking at their curricula vitae. This shows that all scientists are different, and you do not have to be a certain type of person to become a scientist. Interpretive methods like this overcome the high costs of having real scientists in exhibitions, and the Deutsches Museum hopes to use them in our new chemistry gallery if finances do not stretch to the real thing.

Are these techniques so extreme?

While presenting science-in-action in museums might be perceived as a modern and radical way of exhibiting contemporary science, a historical perspective shows that this approach is far from new. In his essay 'Museological Science?', Pickstone points out that analytical science, technology and medicine

not only thrived in museums, but owe their very existence to the new museological institutions that arose from the French Revolution.[16] The prime goal of these museums and quasi-museums was to classify large amounts of information. This was applied to astronomical data, pressed plants and hospital patients alike, and the new sciences that emerged from these collections of specimens reflect the work still undertaken in some natural history museums today. The focus of scientific research moved to the universities only at the end of the nineteenth century, when the work of museums was overshadowed by the experiments that could be performed in university laboratories. Peters believes that some natural history museums have given up their role as up-to-date scientific institutions too easily, and it is only the recent interest in biodiversity that has served to upgrade fundamental research tasks such as stocktaking and taxonomy.[17]

Another category of museums, the fair-like expositions and museums of industrial safety ('social museums'), were dependent on remaining contemporary for their survival. Described by Poser,[18] these institutions illustrated the latest technological developments in safety equipment. They had to be constantly updated in order to fulfil their function as information resources for training, and hygiene and safety awareness. This is in direct contrast to museums such as the Deutsches Museum which were conceived to illustrate up-to-date material, but which gradually changed with time into historical institutions that still serve a purpose in displaying objects of cultural value.

It is instructive to note that the social museums, which appeared throughout Europe during the 1890s, exerted a direct influence on social problems in much the same way that today's science museums hope to achieve their extended educational mission. Bismarck himself saw the *General German Exhibition for the Prevention of Accidents* in 1888 as an important means of demonstrating that the State and industries were concerned for the welfare of their employees. The exhibitions had a function not only in reducing social tensions but also in illustrating how technology could improve the quality of life. They presented technology as the basis of modern culture, and were more successful in promoting its acceptance than the technical museums of the time, which showed only 'masterpieces' of science and technology.

I would like to give two examples of how contemporary research may be successfully demonstrated in today's science museums. The first is the *Hurricane 1724* exhibition at the Museu de la Ciencia in Barcelona.[19] This is a multidisciplinary presentation which tells the story of the sailors and shipwrecks on the Mercury Route. It combines moving exhibits with music and over 300 facts from contemporary underwater research, archaeology and archaeometry. An unusual feature of the exhibition is that it was the Museum's own employees, including its director, who completed more than 40 dives to compile and analyse the evidence. The second example is the educational programme of the Monterey Bay Aquarium in California, developed in 1989. *Live from Monterey Canyon* presents actual video footage from a submersible as it gathers information from deep-sea areas. Visitors can thus observe research as it takes place, and their participation in the project is facilitated by explainers trained in marine biology.[20] Such imaginative presentations of contemporary research are now becoming less of a rarity in today's science museums.

Historical experiments go contemporary

A compromise between presenting historical and contemporary research in museums exists where academic historians of science and technology are moving away from their traditional role of analysing texts and becoming more involved in the museum's work of reconstructing ancient apparatus and instruments. The *Renaissance Engineers* exhibition at the Istituto e Museo di Storia della Scienza, Florence, illustrates this well.[21] However, the Museum has yet to find ways for the visitors to see the machines in action, or even set them into action themselves. At least videos and models are now being considered.

Another approach is to consider the application of contemporary research methodology to historical experiments. It is generally agreed that interpretation of historical scientific research yields the best results when original apparatus is used in a faithful reproduction of the original experiment. However, the use of contemporary methods may reveal new information. For example, replication of a simple chemical experiment originally carried out in London during the 1850s revealed an unexpected structure that was exciting enough by modern standards to achieve publication in *Chemical Communications*.[22] And while this particular experiment will not be described in the new chemistry gallery, the Plinius Project that links several German chemistry faculties with those of mineralogy, physics, philology and archaeology, could well feature.

Contemporary science exhibitions at the Deutsches Museum

While I have considered the Deutsches Museum's popular lectures and links to academia as ways of presenting contemporary science, I have said little so far on the content of the exhibitions themselves. Exhibitions are, according to Peters, 'the most complex way of imparting knowledge',[23] and they constitute the prime activity of the Deutsches Museum.

The original Deutsches Museum is occasionally called the 'old dinosaur of the Isar Island'. We decided to prevent the Museum from living up to its nickname by limiting its growth at that site. Exhibition space of more than 50,000 m^2 and a tour of over 20 km would not be practical, especially in today's society where preferred leisure venues tend to be single sites with easily accessible information. In response to this trend, the Deutsches Museum has metastasised. It opened the Schleißheim Air and Space Museum in 1992 and the Deutsches Museum, Bonn in November 1995. This latest branch of the Museum is remarkable for two fundamental reasons. It is the first museum to focus on research and technology in both East and West Germany after 1945; it is also the first big exhibition to be truly multiperspective, integral-holistic and thematically interrelated.[24] Frühwald, President of the German Research Foundation and keynote speaker at the opening ceremony, described the Museum as taking on the role of editor: it changes dry theory into vivid and interactive exhibits that present the knowledge in an accessible way. He described the Bonn branch as more modest yet more ambitious than its mother foundation in Munich. A good example of how this new museum can respond to the difficult challenge of presenting contemporary science can be found in the exhibition of Nobel prize-winners' work. Less than a week after Professor Christiane Nüßlein-Vollhardt heard that she was to win the Nobel prize for medicine in 1995, and only one week before the Bonn Museum was to open, a large delivery of her laboratory apparatus for studying the fruitfly, *Drosophila*, arrived at the Museum. It was installed in time for the opening, and now resides next to equipment from other prize-winners, such as the pages from Klaus von Klitzing's laboratory notebook, and 90 other marvellous artefacts, all beautifully displayed: masterpieces of science, each with an iconic aura.

The future

The end of the twentieth century has brought with it a tendency for people to hold forth on the future of everything—culture, technology and mankind. As Gunter Hofmann remarked: 'Future, vision, utopia—somehow they all sound like citations . . . behind these terms nests general perplexity.' The Deutsches Museum is attempting to overcome this 'perplexity' with three major projects that will rely almost entirely on presenting contemporary fields of knowledge: the chemistry gallery on the Isar Island site (2500 m^2), the mobility museum at the old fairground site (10,000 m^2)

and a series of high-tech exhibits including a children's museum on the 4000–5000 m^2 made available on the Island by the relocation of the trains and cars. Even though chemistry is expected to become the science of the twenty-first century, the new gallery will include an alchemy laboratory and the historical laboratories of Antoine Laurent Lavoisier and Liebig. For, as Robert Fox pointed out: 'The judicious use of historical exhibits can help to illustrate even the most modern displays of science and technology today.'[25] Thus, the old will mix with the new multimedia displays and interactive exhibits, and real people in the form of explainers, student-scouts, teacher trainees, Humboldt awardees and scientists, in a co-ordinated approach to develop a confidential relationship with the visitor and communicate at least some of the curiosity and fun that science inspires.

Financial limitations restrict the Deutsches Museum's ability to convert the fairground halls built in 1908. Instead, there is a vision for a third- or even fourth-generation museum built by the visitors themselves, and other institutions. The Oktoberfest Museum, automobile clubs, public services and the municipal sewerage system are just a few of the organisations that hope to be involved in the new Deutsches Museum. Workshops, shows and temporary exhibits will fill the space, many of which will focus on telecommunications and media. Contemporary energy technology will be exploited to improve the heating in the poorly insulated halls.

Conclusion

I have described ways in which contemporary science can or already has gained an entrée to science museums. I have considered the strong arguments in favour of this movement, and outlined how the Deutsches Museum presents up-to-date science to the public in an accessible way through programmes such as its popular lectures, its links with universities, and the design and content of its exhibitions.

While the Deutsches Museum has yet to meet the tremendous challenge of filling the 20,000 m² it has at its disposal, one thing is clear: as the twenty-first century approaches, the need to allow individuals to come to terms with the scientific and technological developments of their society is increasingly apparent. Presenting contemporary research in museums is one way of making science more accessible: when it comes to working out the best approach to achieve this, all science museums are in the same boat.

Notes and references

1 Prandtl, W, *Die Geschichte des chemischen Laboratoriums der bayerischen Akademie der Wissenschaften in München* (Weinheim: Verlag Chemie, 1952)

2 Krätz, O, *Historische chemische und physikalische Versuche*, vol. 7 (Köln: Aulis Verlag, 1979), p218f

3 Ebel, G and Lührs, O, 'URANIA—eine Idee, eine Bewegung, eine Institution wird 100 Jahre alt', in *100 Jahre URANIA BERLIN, Festschrift, Wissenschaft heute für morgen* (Berlin: URANIA Berlin e.V., 1988), pp15–74

4 The 1903 Circular of the Provisional Committee, of which O von Miller and W von Dyck were presidents, read as follows:

'. . . in such a museum the great influence of scientific research on technical progress should be illustrated and the historical development of the industry up to the latest achievements shown through outstanding and typical masterpieces. . . to collect many instruments and machines which represent important turning points in the development of modern technologies. Such a museum which comprises all fields of technology would offer the scientific and industrial class a source of constant stimulation and encouragement of their work and would certainly contribute to inspiring future generations and to increasing the glory of our country.'

[*Se. Kgl Hoheit Prinz Ludwig von Bayern*] 'What is the assembly aiming at? The assembly wishes that the masterworks of the present time be collected in a museum . . . the most recent technical inventions which, however, should only be applied once having been checked in practical operation and found reliable.'

5 Radkau, J, 'Zwischen Massenproduktion und Magie—Das Deutsche Museum: Zur Dialektik von Technikmuseen und Technikgeschichte', *Kultur & Technik*, 1 (1992), pp50–58

6 Frühwald, W, ' "Der Herr der Erde—Technikbilder der Moderne oder Die neue Produktion des Wissens", Festvortrag zur Eröffnung des Deutschen Museums Bonn', in Frieß, P and Steiner, P M (eds), *TechnikDialog 3* (Bonn: Deutsches Museum, 1995); MPG Spiegel 2 (München: Generalverwaltung der Max-Planck-Gesellschaft, 1996), pp59–64

7 The Statutes of the Deutsches Museum in 1903–04 stated that: 'The Deutsches Museum serves the purpose of showing the historical development of scientific research, technology and industry interacting with each other, and illustrating their most important steps through outstanding and typical masterpieces . . . .' In 1921 they were changed to read '. . . illustrating their most important steps through *instructive and inspiring presentations, in particular, however, through* outstanding and typical masterpieces . . . .' In 1948–49, the first part was amended, in light of the Museum establishing its own research unit, to read: 'The Deutsches Museum serves the purpose of *investigating* the historical development of science, technology and industry, to show *their interaction* and to illustrate . . . .' A further alteration was made in 1976: ' . . . to illustrate their most important steps through *cultural importance*, instructive and inspiring presentations, in

particular, however, through outstanding and typical masterpieces . . ..'

8 I believe that this can be concluded from the recent critical summary of modern literature by the writer Andreas Neumeister, 1996 winner of the Bavarian national prize for literature. He deplores the lack of contemporaneity in contemporary literature: Neumeister, A, '50 Jahre Nachkriegselend', *Süddeutsche Zeitung*, 17 October 1996.

9 Minutes of the CEFIC-ICASE Conference, Dresden, 20–21 November 1994

10 European Commission, 'Teaching and learning—towards the learning society', *White Paper* (Luxembourg: European Commission, 1995)

11 Hantke, B and Schärer-Züblin, E, this volume, pp263–70

12 Schmidt, H and Graßl, H, 'Die Bringschuld der Wissenschaft und die Annahmepflicht der Politiker', in Frieß, P and Steiner, P M (eds), *TechnikDialog 3*

13 Graf, B and Treinen, H, 'Besucher im technischen Museum: zum Besucherverhalten im Deutschen Museum München', *Berliner Schriften zur Museumskunde*, 4 (1983); Wersig, G, Professor, FU Berlin, private communication (1994)

14 Miles, R S, 'Museums and the communication of science', *Communicating Science to the Public* (Chichester: Wiley, 1987), pp114–22

15 Wengenroth, U, 'Kulturgeschichte und Technik im Museum', *Siemens Kulturprogramm 1989/90* (München: Siemens AG, 1990), pp64–65; Wengenroth, U, 'Zur Geschichte des Vertrauens in Wissenschaft und Technik—Menschenbilder in der Technik und die säkularisierte Heilsbedürftigkeit', in Weis, K (ed), *Bilder vom Menschen in Wissenschaft, Technik und Religion* (München: Technische Universität, 1993), pp163–87

16 Pickstone, J V, 'Museological science? The place of the analytical/comparative in nineteenth-century science, technology and medicine', *History of Science*, 32 (1994), pp111–38

17 Peters, D S, 'Das Naturkundemuseum als Ort der Forschung und Wissensvermittlung', *Tendenzen 95* (Bremen: Überseemuseum, 1995), pp59–66

18 Poser, S, 'Museum of Danger—Dangerous Museums?', *Proceedings of the 23rd Symposium of ICOTHEC*, Budapest, 7–11 August 1996, in press

19 Wagensberg, J, 'Hurricane 1724', *ECSITE Newsletter*, 26 (1996), p1

20 Connor, J L, 'Promoting deeper interest in science', *Curator*, 34 (1991), pp245–60

21 Galluzzi, P, *Gli ingegneri del Rinascimento. Da Brunelleschi a Leonardo* (Firenze: Istituto e Museo di Storia della Scienza, 1995)

22 Rieger, D, Hahn, F E and Fehlhammer, W P, 'Chemistry of hydrogen isocyanide, Part 6. The supercomplex nature of Buff's ferrocyanäthyl', *Chemical Communications* (1990), pp285–86

23 Peters, D S

24 These attributes are characteristic of the Mode 2 production of knowledge described in Gibbons, M, Limoges, C, Nowotny, H *et al.*, *The New Production of Knowledge. The Dynamics of Science and Research in Contemporary Societies* (London: Sage Publications, 1994).

25 Fox, R, Professor, Oxford University, personal communication (1995)

Science on television: a coming of age?

Jana Bennett

Science on British television has changed radically over the last 30 years. Can museums learn from the challenges that programme makers face?

What view of contemporary science and technology do the mass media present to the public? This paper is confined to the medium of television, not only because this is the main medium in which I have worked, but because television is probably the largest source of information on contemporary science used by the general public outside formal education. I would not claim that television can offer science museums and science centres a way forward as they seek to increase their coverage of topical science stories and new developments. But, as a programme maker I would argue that museum professionals will draw some interesting parallels with the changing nature of contemporary science programming over the past 30 years. Exhibition developers may find some useful information in the research we have used to shape our current approach. If museums and science centres are to begin to tackle subjects and issues which up to now have only been covered in other media, then I suspect we may face some similar challenges.

However, a warning about evidence: the television industry is not very sophisticated in terms of its qualitative research methods, and available data on the content of science programming and audience responses to it are not conclusive. However, the data that exist and the quantitative research that measures viewer numbers, demographics and viewing habits are both revealing and robust.

One common challenge which is increasingly facing both the museum world and that of television is the search for audiences—for 'visitors'. In both museums and increasingly television there is a need to recruit our audiences in competition with the plethora of other possible activities, from video games, theme parks, the World Wide Web and, in the BBC's case, from cable, satellite and shortly from hundreds of competing stations transmitting via digital technology. We are in a world of fewer and fewer captive audiences. Audiences are also seeing their time as a form of leisure expenditure which they will decide how to broker for themselves. They will make the choice between theme park and museum, between a BBC programme and a computer-games console. The media will have to focus clearly on what it can and cannot do in order to provide something attractive to them. By identifying some of these trends, the BBC has become more audience-focused and less paternalistic in its science programme provision. Given that, as programme makers, we have less and less influence over what people watch, this seems to be the right focus.

There is another common force operating for both museums and television: the interest that people have in knowing more about science and technology. The link between science, technology and industrial expansion is encouraging many populations to become more technologically literate. People in other countries may be more aware than those in Britain of the importance of understanding the impact of scientific developments.[1] Thus it is no accident that the BBC is continuing to expand the coverage of science on World Service Radio through its news and features coverage. This expansion is continuing on television channels, not just in the US and the UK but elsewhere in the world.

I have painted a rather optimistic picture of infinite thirst for scientific and technological information. A subsequent question is:

demand for what? It is important to ask what approach to science and technology is needed or indeed wanted. Is the depiction of science by the media utopian or does it adopt a critical approach?

A short history of British television's science coverage

Science on television grew out of the fact that television itself is a technological wonder. Sixty years ago the first BBC television broadcasts merely screened what they found in the real world or, more often than not, what they found in the Alexander Palace studios. Even 33 years ago the original mission of the flagship *Horizon* programme, 'The World of Buckminster Fuller', transmitted on BBC 2 on 2 May 1964, was to translate the ideas of contemporary science on to the screen for a lay audience and to act as an interpreter for the scientists:

The aim of *Horizon* is to provide a platform from which some of the world's greatest scientists and philosophers can communicate their curiosity, observations and reflections, and infuse into our common knowledge their changing views of the universe. We shall do this by presenting science not as a series of isolated discoveries but as a continuing growth of thought, a philosophy which is an essential part of our twentieth-century culture.[2]

That was, of course, only for those scientists who were vulgar enough to use the medium at all. There have been many stories of how scientists were spurned by their colleagues for having 'supped with the television devil'. Even now it is sometimes considered harmful for a scientist to have appeared 'on the box'. From Professor Jacob Bronowski onwards, 'telly boffins' have sometimes had a bad time back at the lab. Bronowski, presenter of the renowned *The Ascent of Man*, felt he was held in less high regard as an academic because of his role in popularising (vulgarising) science. However, this mode of passive translation coupled with the broadcaster's initial deference to science ensured that the first television

science broadcasts presented an optimistic view of the future which could justifiably be called utopian. An episode of *Horizon* from 1964 illustrates this attitude of awe and belief in the future being built by science and technology (excerpt 1).

This typifies how modern science on television started out with a utopian but also a genuflecting outlook. The programme makers were granting the viewers the privilege of an audience with the scientist on the screen. This left little room for debate, critique or comment. However, it is important to stress that the treatment of politics was much the same on television at this time.[3]

Science on television has been very much a product of the age. Consequently, television was swept along by the social upheavals in the 1960s, and so the critical social documentary was born. English science programmes were no different from the rest of the media, or indeed the rest of society. Producers were not doing their job if they were simply passive translators. Society at large was taking on big issues and by the late 1960s science television was following suit. It wanted to have a point of view. Anthony Wedgwood Benn, the incoming Labour government's Minister of Technology, was invited to present a whole programme discussing the impact of technology on society.[4] Campaigning journalism extended to television, where the latest science became a provider of evidence, used to attack or expose targets that ranged from the dangers of blue asbestos and tobacco, to people's tendency to obey authority. This shift in attitude represents neither a negative nor a utopian view, but is a recognition of the power of science to do good or evil. Television producers also utilised science information themselves to pursue an issue. In a famous *Horizon* 'You Do as You Are Told' programme,[5] audiences watched stunned as ordinary volunteers administered ever more intense electric shocks to a recalcitrant subject who was being told that they must obey. When they didn't obey they had to be punished. Only later did viewers, and indeed the punishers, discover that the

Excerpt 1. Programme *Horizon*: 'The Knowledge Explosion', transmission 21 September 1964 on BBC 2, producer Michael Lathamboth

Narrator What remains for science to do? How will it affect our lives? This is a city of the near future, planned by scientists and designers for the General Motors exhibit at the World's Fair in New York. They see a future where man will be making fuller use of the world's at present un-

tapped resources. Improved technology will make it possible to penetrate jungles and build roads with tools so efficient that from tree cutting with a laser beam to laying road foundations will be a matter of only a few short hours with equipment like this. In this world of tomorrow overland communications will be vital, say the experts at the World's Fair. So super highways will cross areas which are now served only by cart tracks.

Arthur C Clarke The only thing we can be sure of about the future is that it will be absolutely fantastic.

near-lethal voltages were merely buzzers making a noise and that the person on the other side of the screen was an actor. The programme demonstrated how far psychological techniques could be used to make somebody into someone who would just take orders. This was at a time of great concern about brainwashing, about the Vietnam war, about the military industrial complex—all with a fair degree of paranoia thrown in. 'You Do as You Are Told' and other programmes marked an increased suspicion about contemporary science in the late 1960s and through much of the 1970s: what was science being used for?

By the 1980s media attention moved to the politics of science itself: how the scientific community functioned as an industry in its own right. 'Star Wars' was the subject of a half-hour special devoted to the politics and technology of the space-based strategic defence initiative on the prime-time magazine show *Tomorrow's World*.[6] HIV and AIDS received a similar treatment,[7] as did Chernobyl on BBC 1.[8] The Chernobyl programme was presented from the studio and, rather

dangerously it seemed at the time, live from the lid of a nuclear reactor.

A comparison of three programmes on genetic screening and its potential uses outlines three different approaches and indicates where science broadcasting is heading. 'Brave New Babies' (excerpt 2) was produced by *Horizon* in the 1980s. This dramatisation shows two parents selecting their future child's physical make-up and character with a medical technician who enters their requirements into a computer. The next transcript is from the *Antenna* magazine programme in the early 1990s (excerpt 3). It dramatised a situation in which the health insurance salesman of the future is doing a genetic profile of his client, a rather anxious young man. The third is from a 1996 broadcast, the BBC 2 series *In the Blood* (excerpt 4). The series about genetics had a more down-to-earth approach to genetic screening— in this case showing how a family deals with a rare and inherited cancer. The 1996 television treatment of genetic issues is no longer tinged with suspicion, but is practical. In this extract, a mother of two

Excerpt 2. Programme *Horizon*: 'Brave New Babies', transmission 15 November 1982 on BBC 2, producers David Dugan and Oliver Morse

Medical technician You've had a chance to view the data at home?

Mother Yes. We've narrowed it down to zygotes three and six. We're not really sure which one to choose.

Medical technician What sort of characteristics were you thinking of?

Father Well we definitely don't want to tamper with the physical side of things in any way.

Mother No, except that we would like her to have my father's red hair.

Medical Technician Ah! Oh well that's easy. We can make her homozygous on the three hair-colour genes. What about her character and emotions?

Mother Ah, well yes there are a few things we'd like to have modified if possible. We'd like to reduce shyness, and susceptibility to depression, without necessarily damaging any artistic potential. Also we'd like her to be musical and if possible, also we want her to be ambitious.

Father Of course we want her to be as healthy as possible.

A medical technician identifies parents' requirements for their forthcoming child in 'Brave New Babies'

Excerpt 3. Programme *Antenna*: 'Nobody's Perfect', transmission 16 January 1991 on BBC 2, producer Tim Haines

Insurance salesman OK Jose! Let's see what the men in white coats made of you.

Client What are you looking for these days? I mean, there seems to be more and more added each year.

Insurance salesman Only the main 'baddies'. Oh yeah we have a very strict code on that one. Only those approved by the committee. Heart, mind, lungs, stomach. Yeah. Pretty much everything, but the more we have the more we can tailor the policy to your needs. And of course, all these juicy facts about you John are for my eyes only.

Client Yeah, I bet.

Insurance salesman Good, good. Well . . .

Excerpt 4. Programme *In the Blood*: 'The End of Evolution', transmission 24 June 1996 on BBC 2, producers Robin Brightwell and Dana Purvis

Mother I don't have worries or concerns about myself because I go for regular tests, but I do worry about the girls obviously because I'm a parent and they're my children. I worry if I have passed something on.

Father This is one of the things that we discussed when we decided to have children: it could be anything. It's not as if we think it's a special case, it's just part of life, and a lot of people will be affected by various genetic diseases.

Mother Hopefully, by the time they're due to be tested—round about 13 or 14, you know when they reach adolescence, they'll be able to find out from blood tests so that they don't have to go through the examinations that I did.

Steve Jones What would your advice be to other people who might be dubious about having genetic tests?

Mother I would have them.

Father If Deborah hadn't had any of the tests she wouldn't have been here now. So obviously if there is a risk of something, then you have got to be tested and if things are caught early enough, then modern medicine can get it sorted out.

Steve Jones with two children featured in the series In the Blood, *on modern genetics*

Professor John Burn People come along and say 'Have I got the gene for this because my dad died of it?', and the answer is I need your dad's blood sample to answer the question because I need to know which of the thousands of possible spelling mistakes actually caused the disease in your father. You can get round it to some extent, but if we've stored samples from affected family members it makes an enormous difference and that's what we are doing now. We've got something like 16,000 people's DNA stored away for a rainy day.

Steve Jones And how many will you have in 20 years?

Professor John Burn I hate to think, but the power bill could be enormous.

young children talks to Professor Steve Jones, a geneticist, about genetic testing of her daughters for an inheritable bowel cancer. These programmes show a progression from a fearful attitude towards genetic engineering, using dramatic futuristic scenarios in the two earlier examples, to a more immediate and information-rich perspective as the science comes closer to home, as portrayed by *In the Blood*. Although part of this change may be because genetic screening is a reality today and not just some prophecy belonging to a future world, I would suggest that the media's attitude has also shifted from one of fear to one of 'let us assess and debate'. *In the Blood* aimed to cover the public debate about the possible dangers of genetic screening, yet avoided scare tactics. The *Horizon* 'Brave New Babies' approach of using drama or satire to make a futuristic point is less likely to be used today by science programme makers. Perhaps it's because a more mature relationship is developing between the media and science. Whether television and society are well prepared to deal with some of the stranger effects of medical technology such as selective termination of pregnancy and cloning is, however, still questionable.

Science television grows up

I have suggested that there may be a new, more mature relationship between the science community and the media. There is, however, another factor: the increased need for the media to understand and respond to audiences' interests and likes.

In an analysis of news bulletins from February 1994 conducted by the BBC,[9] researchers were interested in what provides a news item's appeal, and what prevents it being effective as far as viewers were concerned. The survey established that there is potentially more interest in science stories in the news than in coverage of the arts, sports, finance or party politics. Genetics, medicine, environmental issues and other science stories which have relevance to people's lives aroused particular interest. Importantly, for a science story to qualify as newsworthy, the ordinary viewer has to be able to understand it. This creates problems for coverage of modern science.

Likely to fail vs likely to succeed

Table 1 highlights the difficulties caused by the very nature of science as a discipline. Much of science is based on claims and hypotheses, and real milestones and achievements are sparse. Much of what

Table 1. Characteristics associated with science items in the news which indicate whether a programme is likely to succeed or fail

Likely to fail

- 'Science-for-science's-sake'
- No relevance to everyday life
- Doesn't indicate *why* we should be interested
- 'Claim'/hypothesis
- Viewer knows nothing about area
- Complex/difficult/technical/concentrates on *how* research is done
- Footage of boffins/machinery
- Scientific/technical jargon
- Long/rambling
- Covers many areas
- Gives publicity to vested interests/shows bias
- Ignores viewers' worries

Likely to succeed

- 'Science-for-the-human-race'
- Could affect us all
- Indicates *why* we should be interested/why we are being told *now*
- Fact/real achievement/milestone
- Viewer already knows enough to be able to integrate new information
- Presented simply, without too much explanation of technical/theoretical background
- Clear, explanatory graphics
- Layman's language
- Short
- Focuses on one clear issue
- Performs public service (warns of danger/flags where help available)
- Shows awareness of viewers' concerns

science is about is painstaking testing and uncertainty, yet this is something which, if concentrated on, is likely to make a 'story' fail. The complex, the difficult, the technical and the concentration on 'the how' are essential to actually understanding what a scientific development really means. Science for science's sake is very important within science, and many scientists rightly feel passionately about this, but it often leaves television viewers cold. In contrast, what is likely to succeed is media-hyped science, claiming a real milestone, real certainty and achievement. Take the case of the 1996 Nobel prize winner Sir Harry Kroto's discovery of buckminsterfullerenes or 'bucky balls'. To say this new form of carbon is directly relevant to our lives would be an exaggeration, and yet this was an

exciting science story. There are, therefore, some aspects of science itself which do not lend themselves to successful science communication.

Newsworthiness of science items

Table 2 shows 23 science stories rated by focus groups according to newsworthiness. What comes across is that the clear key to the more newsworthy stories is relevance. One respondent said about 'Missing Matter', a clear

Table 2. Number of focus groups rating a science item as 'newsworthy'

Science items	Yes	Don't know	No
1 Government go-ahead for genetically altered food	8		
2 Scientific advances in genetic screening raise serious moral concerns	8		
3 Nuclear reprocessing: government gives go-ahead for Sellafield tests	8		
4 Trials begin of a new treatment for breast cancer	8		
5 Americans clone identical twins	8		
6 British scientist receives Nobel prize for medicine	8		
7 Gene therapy gets the official go-ahead	7		1
8 A thousand women in the West Midlands recalled after examinations for cervical smear tests	7	1	
9 Nuclear fusion: American scientists claim major advance	6	2	
10 Ice-core research throws doubt on global warming	6	2	
11 Hubble telescope: astronauts replace faulty parts	6		2
12 Experts say there's no evidence to link children's skull deformities with polluting chemicals—but North Yorkshire parents demand full investigation	6		2
13 AIDS: Department of Health report predicts sharp rise in number of British heterosexuals contracting HIV virus	5	3	
14 US researchers think they have located 'gay gene'—but worry that it may lead to sex-orientation checks on the unborn	5	2	1
15 A new British clinic offers sex selection of your baby	5	1	2
16 Trials of anti-AIDS drug AZT show that it does not delay the onset of the disease	4	4	
17 President Clinton announces plans for new US Space Station	4	2	2
18 Government reveals plans for shake-up of science and technology	3	4	1
19 Video games are turning many young people into addicts and making them aggressive	3	3	2
20 New telephone technology paves the way for home banking	3	3	2
21 Computer scientists warn that British lead in technology could be lost	2	2	4
22 Australian scientists find universe's missing matter	1	4	3
23 British scientists race towards absolute zero	1	1	6

57

'loser': 'Why should I care? I didn't even know it was missing.' An item entitled 'British Scientists Race towards Absolute Zero' fell to the bottom of the league table because the story appeared to rely upon science for science's sake. The analysis of why stories fail reveals that the viewing public has a sophisticated screening system for relevance.

Programme makers have to take account of this negative response to pure science when thinking of how to cover scientific news: rare diseases, foreign achievements, continuing projects without new milestones or with no solution in sight, government shake-ups, a lead in technology that is likely to be lost, or raising false hopes such as whether cancer treatment is promising too much—this is a long list of 'loser stories'. I believe programme makers should be covering these issues—although with care—even if they are not considered to be immediately high up in news agenda or even on the news agenda.

What is needed is a filter of relevance, or a mechanism for constructing that relevance if necessary through the way the story or subject is presented. For instance *Horizon:* 'Assault on the Male' made specialist science relevant for the non-news-programme audience, yet was presented in the form of an extremely news-worthy 'scoop' documentary, as is illustrated in excerpt 5. This was the first time any broad-caster had gathered together the news about new oestrogens in the environment.

Even though there is quite a lot of science and chemistry in this programme, this point of relevance, the abnormal alligator genitalia, was never exaggerated but always driven home. This made a very successful programme and one which has exposed the subject to a lot of public debate.

A second lesson is that of following a process. Involving audiences in the conduct of contemporary science enables difficult and abstract science topics to work for audiences—especially for longer reports such as documentaries of more than 30 minutes in length. Seeing science as a process, as method in action, also links science as an activity to other areas of human endeavour. *Horizon:* 'Ulcer Wars' is an example which combines process with relevance and hence subsequently in-volves the audience. This extract illustrates an unusually direct way of testing an unusual hypothesis: that ulcers could be caused by a bacterium, *Helicobacter pylori*, which, if true, could theoretically be cured with antibiotics. The test was unusual because it was first tried out by the friends of the scientist involved, Barry Marshall (excerpt 6).

This was an example of a patient with *Helicobacter pylori* bacterium, the discovery of which was overturning the prevailing stomach-ulcer hypothesis. It proved to be a contentious issue, as the drugs industry had not really moved with theory. The point emphasised in the programme was that a scientist, Barry Marshall, was ahead of the pharmaceutical industry's ideas and was ignored because he had an unconventional idea which also chal-lenged a hugely lucrative industry. Such ap-proaches to science in television are basically journalistic, although not in the traditional way of news. This does not represent a utopian or critical vision, but a practical view of science which assesses the current standing of scien-tific ideas.

Horizon concentrates on process, on rel-evance, on stories from the world of science with occasional uncertainty and human fail-ings thrown in. Yet, while wanting to be con-temporary and timely, it does not always strive to be topical. The science comes across as an area of rolling human activity with few points of absolute discovery, while the highlighted moments may or may not stand for 'good' progress.

'Good' vs 'bad' science

What effects do the good news or bad news messages within programmes have upon viewers? Assessing *Horizon* by topic using audience appreciation scores[10] provides an interesting perspective on its 33-year history. In the audience appreciation index, a high score is in the high 70s to the mid 80s and a

Excerpt 5. Programme *Horizon*: 'Assault on the Male', transmission 31 October 1993 on BBC 2, producer Deborah Cadbury

Narrator Lake Apopca, Central Florida. A scientific team was called in to investigate the declining number of alligators. They found more than they bargained for.

Professor Louis Guillette We were astonished at what we found. Things that we were seeing were so dramatic. We were actually seeing sex reversal. I mean things were changing sex. At least 25 per cent if not 30 per cent of the male alligators on this lake have some kind of abnormal phallus or abnormal penis. Mostly it appears that the abnormality is small size, as much as a half or two-thirds reduced. For this size animal, it's normally about twice this wide at least. But it doesn't ever have this hook to it like this.

Narrator Similar changes to males of other species have been found in Europe and America and now there are signs that human reproduction could be in trouble.

Professor Louis Guillette Everything that we are seeing in wildlife has an implication for humans. I believe that we have the potential to have major human reproduction failures.

Narrator These are human sperm. They show a high level of abnormality. These have grossly deformed heads. Whereas these have no tail, this one is all tail. And this sperm has two tails. Here the neck is enlarged with cytoplasm which should have been shed earlier. And while some are hyperactive, others don't move at all. It was these sorts of abnormalities which scientists from Copenhagen began to study two years ago. At the University Hospital, Professor Skakkebaek was surprised to find some healthy normal men had more than 50 per cent abnormal sperm.

Louis Guillette and Timothy Gross on an expedition in Florida's Everglades to investigate the declining number of alligators. They find strange things lurking in the swamp. Are males under threat?

Excerpt 6. Programme *Horizon*: 'Ulcer Wars', transmission 16 May 1994 on BBC 2, producer Michael Mosley

Narrator One of the first to try his antibiotic cocktail was Win Warren, his colleague Robin Warren's wife.

Win Warren Once it was clear that I was one of the folk who had these bugs growing in my stomach, it was arranged that I should go to Freemantle Hospital for gastroscopy which I didn't find an entirely delightful procedure. There was lots of grizzling, from Barry in particular, about the fact that I kept chewing on his gastroscope, which I was informed cost several thousand dollars and had to be treated with respect. I'd have liked a little respect myself at the time. So that bit of a treat, swallowing the gastroscope, was followed by great excitement on Barry's part because low and behold there was the ulcer. I could have lived without sharing his enthusiasm because he insisted that I looked down the eyepiece. There's something disgusting and obscene about looking at the inside of one's own stomach but he seemed to think it was the most exciting thing he'd seen for a long time. After that it was very clear that I was part of the trials as far as taking medication was concerned, and that was a relatively simple thing

Barry Marshall looks right into our stomachs to wage 'Ulcer Wars'

to do. There were no major side effects of the pills, and before very long I was nicely cured and eating well again. That was nearly ten years ago and I've had no recurrence whatsoever. Sometimes when my weight goes up too much, I think maybe a recurrence wouldn't be such a bad thing but I think the likelihood is nil.

lower score is the mid 70s. Contemporary topics with worrying secondary issues (AIDS or the non-ethical introduction of Norplant, the implantable female contraceptive, into developing countries, for example), received lower scores in the audience appreciation indices than 'discovery' programmes. The *Horizon* programme 'AIDS: Behind Closed Doors',[11] for example, received a score of 75 and 'The Human Laboratory'[12] (the Norplant story) scored 76. A more popular story called 'The Planet Hunters',[13] about discovering a new planet, recently scored 80.

On the same scale, a programme on the ethics of non-lethal weapons and the politics of land mines, *Horizon*: 'Small Arms, Soft Targets'[14] received a score of only 74. Socio-political issues and campaigning films seem to be given relatively low scores by the audience, unless they have empowering and optimistic endings. My interpretation of the audience appreciation scores suggests that the audience is searching for hopeful messages: for stories from the world of science which are enriching and perhaps complicated, yet which do not give prominence to worrying issues.

An important exception to this rough-and-ready rule was coverage of bovine spongiform encephalopathy (BSE), a serious subject which was ranked highly in the audience appreciation index. This is a particularly British contemporary issue about which viewers are worried and personally interested. Seventy-four per cent vs 21 per cent of viewers in our focus-group research stated clear preferences for narrative rather than 'fact-file' approaches. This supported the approach taken by the producer of the recent *Horizon* programmes examining BSE, illustrated in excerpt 7, which focused on a strong personal story within the wider scientific framework. The programme covered prions, public policy and scientific uncertainty. The point about uncertainty was that in the case of CJD we do not know where we are on any possible epidemic curve. Are we at the beginning of a large curve, or will there be only a few cases? The programme discussed epidemiology and how

---

Excerpt 7. Programme *Horizon:* 'The Human Experiment', transmission 18 November 1996, producer Bettina Lerner

A surgeon changes into clothes preparing to begin a Creutzfeldt–Jakob disease (CJD) autopsy.

Narrator In 1990 the government set up a small research unit with one sole purpose—to detect the earliest signs of a human epidemic of mad cow disease. They were to study every case of CJD in Britain, to look for any change in the normal pattern of disease, a new variant that would reveal that BSE had crossed into humans. One member of the team had the job of examining the brains of everyone in Britain who died of CJD, looking for anything unusual. But a CJD autopsy is unlike any other a neuropathologist ever has to perform.

Dr James Ironside I've now changed into the clothes I'll wear to do an autopsy in a case of suspected CJD, and these garments are disposable because they will be incinerated after the autopsy. These are the gloves. On the top is a chain-mail hand piece. And this is flexible and allows my hand to be protected from any cuts while the autopsy is being performed, and I wear another pair of rubber gloves on top of this just to make the whole thing as waterproof as possible. And this is the helmet. I'll just put it on. The instruments that I use in the post-mortem room and the instruments that technicians use in our dedicated laboratory—really we regard these as being permanently contaminated, so we use those for CJD cases alone because there is no effective way of guaranteeing decontamination in this disease. Unlike other viruses and bacteria, it can resist extremes of heat, cold, chemicals, enzymes—everything practically, and I don't think any of us believe that we can fully decontaminate anything with this agent.

James Ironside preparing to perform an autopsy on a suspected CJD victim

epidemics can or cannot be identified, a debate which will obviously be important in terms of future health policy. What stance does such a narrative approach take to science? In the case of BSE, the most important question is how the developments in a particular field may affect public policy. Science is not exactly value-free, but is reported as an activity, much as we report on politics, the arts or industry.

The future

This mix of narrative and reporting was reflected in a recent science fiction series that the BBC made, *Future Fantastic* (excerpt 8).[15] This incorporated wonder, yet portrayed visions of the future that had the potential to be good, bad or just plain disturbing.

The programme asked whether 'change' would be equivalent to progress. This was left very much as an open question and was an attitude which the producers believe worked very well for a young audience. *Future Fantastic* attracted more younger people in the 16- to 24-year-old age groups than is usual for a BBC 1 prime-time factual programme.

The contrast of *Future Fantastic* to the first extract on the future is interesting in that it suggests our that today's visions belong both to science fiction and to science. The programme is saying that these visions are not necessarily utopian, but they are subject to reporting based on what is happening today.

Conclusion

There are challenges ahead for science communication, both in television and other media such as exhibitions. The audience is curious and has a sense of wonder, and this needs to be tapped. As Professor Richard Dawkins

---

Excerpt 8. Programme *Future Fantastic Promo,* series transmission 21 July to 30 August 1996 on BBC 1, series producer David McNab

Narrator 'Future Fantastic' reveals over a century of dreams. And how scientists past and present have turned those dreams into reality.

Professor Charles Vacanti Any organ that you can think of that fails in the human body has a potential to become a tissue-engineered organ. I believe we're ready to start doing it in humans now.

Narrator This major new series unveils the startling truth about our future. Can tomorrow's scientists deliver today's science fiction? Are we ever going to conquer space or make contact with aliens? Will our children live on Mars? Will we achieve eternal health and beauty? Will we see advanced bionic people or regrow lost limbs? Teleport around the world or even travel in time? Could today's science-fiction visionaries actually have got our future right? 'Future Fantastic'—it's nearer than you think.

Gillian Anderson, star of US science-fiction drama The X-Files, *took viewers into the* Future Fantastic

---

said in the BBC's 1996 *Richard Dimbleby Lecture*:

The popularity of the paranormal, oddly enough, might even be grounds for encouragement. I think that the appetite for mystery, the enthusiasm for that which we do not understand, is healthy and to be fostered. It is the same appetite which drives the best of true science, and it is an appetite which true science is best qualified to satisfy. Perhaps it is this appetite that underlies the ratings success of the paranormalists.[16]

At the same time, there is a huge interest in the paranormal and it is an immense challenge to establish where science communicators should be placing themselves in terms of either tapping into or ignoring that interest.

There is also a great need to continue to look at how programmes can be relevant and how broadcasters can drive that relevance home. The audience should be compelled to participate, whether by evoking the notion of process in a film, the narrative of a story or, increasingly, by interactivity in the form of Web sites.

Exemplifying this is the *Horizon* Web site,[17] which allowed viewers to follow an expedition to find a Peruvian mummy. Despatches from television cameras were sent to the Web site and on to our international site shared with the Public Broadcasting Service network in America. Viewers were able to participate in the actual expedition they would see on the *Horizon* programme 'Ice Mummies: Frozen in Heaven'[18] some weeks later. The site gave an opportunity to the public to write to the members of the expedition and to the scientists in Peru, an interactivity which created a new interface for the audience.

As I have shown, the presentation of contemporary science on British television has changed during the past 30 years. While it was once possible to criticise science programming as deferential and portraying a universally positive view of science, this is no longer the case. In the 1990s, some of the key features of science coverage on television are the search for relevance, a focus on the impact of breakthroughs and discoveries; and a concentration on process and the narrative when covering science in documentary format. Today, in general, science and technology is treated both as a part of people's daily lives and as a part of industry. If science museums and science centres are to follow television in communicating the science that is in the news, then these features should be assessed for their potential to translate into successful exhibitions.

Finally, it is worth considering whether science is now taken for granted by television—have we as programme makers lost our sense of wonder, and could we be in danger of underrating the power of science by taking a rather cooler look than perhaps in the past? The answer is 'not necessarily'. Broadcasters need to preserve their ability to give a sense of excitement and to reflect on the major developments and genuine discoveries that science offers up. In so doing, we will be able to help distinguish the enormous force for change—for progress—contained within the world of modern science.

Notes and references

1   *European Report on Science and Technology Indicators* (Luxembourg: European Community Publications, 1994) quoted by Robert May, UK Government Chief Scientific Adviser, Office of Science and Technology, London.

2   Daly, P, 'Horizon', *Radio Times* (30 April 1964)

3   For example, Robin Day was acknowledged to be the first severe questioner of politicians in the late 1950s.

4   *Horizon*: 'Machines and People' transmitted on BBC 2 on 5 June 1969, producer Robert Vas

5   *Horizon*: 'You Do as You Are Told' transmitted on BBC 2 on 28 October 1974, producer Christopher La Fontaine

6   *Tomorrow's World Special:* 'Star Wars' transmitted on BBC 1 on 14 November 1985, producers Martin Freeth and Martin Hughes Games

7   *Tomorrow's World* transmitted on BBC 1 on 1 December 1988, producer Martin Mortimore

8   *After Chernobyl—Our Nuclear Future?* transmitted on BBC 1 on 27 May 1986, producer Philip Harding

9   BBC Broadcasting Research, *Science in the News—Qualitative Research SP93/98/3125*

10 BBC Broadcasting Research, *Horizon: Winter 1996 Series Broadcasters' Audience Research Board (BARB) Analysis 6144TV.* Audience appreciation indices (AI) measure how much a sample of viewers have enjoyed or appreciated a programme. Their response is measured immediately after a programme has been transmitted. Respondents' opinions are used to calculate an AI for every BBC, ITV and Channel Four Television programme. The AI ranges from 0 to 100, a high AI indicating a high level of appreciation. AIs are calculated as simple averages of the scores given by respondents. The AI is based on the number of panel members who rated the programme each week. In this respect a panel member who returns a diary is included in the analysis even if he/she has not watched any programmes during the week. For every programme, the number of viewers who marked each position on the 11-point scale is determined. The scale positions 10, 9, 8, 7, 6, 5, 4, 3, 2, 1, 0 are then treated as, respectively, 'scores' of 100, 90, 80, 70, 60, 50, 40, 30, 20, 10, 0, and an average is taken.

11 *Horizon:* 'AIDS—Behind Closed Doors' transmitted on BBC 2 on 4 December 1995, producer Andrew Chitty

12 *Horizon:* 'The Human Laboratory' transmitted on BBC 2 on 5 November 1995, producer Deborah Cadbury

13 *Horizon:* 'Planet Hunters' transmitted on BBC 2 on 11 March 1996, producer Danielle Peck

14 *Horizon:* 'Small Arms, Soft Targets' transmitted on BBC 2 on 10 January 1994, producer Martin Freeth

15 *Future Fantastic* was a nine-part series transmitted weekly on BBC 1 from 21 June to 19 July 1996 and from 9 August to 30 August 1996, series producer David McNab.

16 *Richard Dimbleby Lecture: Science, Delusion and the Appetite for Wonder* by Richard Dawkins transmitted on BBC 1 on 12 November 1996, producer Charles Miller

17 http://www/pbs.org/wgbh/pages/nova/peru/index.html

18 *Horizon:* 'Ice Mummies—a Three-Part Special', transmitted on BBC 2, producer Tim Haines. **1** 'The Ice Maiden' transmitted on 30 January 1997, **2** 'A Life in Ice' transmitted on 6 February 1997, **3** 'Frozen in Heaven' transmitted on 13 February 1997.

Programming for success

Neutrinos in a science centre, part I— view from a programme developer

Munkith Al-Najjar

No subject is too abstruse for a science centre to tackle. Science North is interesting its visitors in neutrinos, the most elusive of fundamental particles.

For most of us, the sub-atomic domain is a distant foreign country. It comprises fundamental particles that are completely outside our everyday experience and which behave in strange ways, according to the laws of quantum mechanics. It is hard enough for university students to come to terms with this unfamiliar territory, so how can we hope to make it interesting to non-specialist visitors to a science centre?

Over the past few years, Science North Centre in Northern Ontario, Canada, has been attempting to acquaint its visitors with the most elusive of all fundamental particles: neutrinos. These tiny particles are currently believed to have neither shape nor size, to have little or no mass, to have no electric charge and to interact only weakly and via gravity.[1] Every second, billions of neutrinos shoot through every human body—having come from the Sun, outer space, and the core of the Earth—but their interactions are so feeble that we are not aware of them.

Why, then, is Science North attempting to interest its visitors in particles as unfamiliar as these? The short answer is that one of the world's leading neutrino detectors, the Sudbury Neutrino Observatory, is being built in our own back yard. It is probably unfair to characterise the Observatory, which is due to be completed in 1997, as contemporary science. A more appropriate term is that it is 'science in the making'.[2] From the beginning of the construction project, the Observatory and Science North recognised the importance of collaboration. The Observatory is interested in establishing links with the community, in providing the public with the latest information about the progress of the project, and in helping to present educational programmes for local school children. For Science North, the collaboration is important as it enables us to fulfil our mission of providing relevant science to the people in our community. As I shall describe, the fruits of this collaboration are a host of interactive exhibits, video presentations, a multimedia theatre and a variety of educational programmes.

Science North

Since opening in 1984, Science North has developed a unique style of science communication. This is accomplished by involving visitors with real science and using actual scientific tools to experiment and explore. The science experience at Science North comes alive with the interaction between the visitor and the science staff who have an in-depth knowledge of science and a passion for learning. The medium of interaction is varied. Low-cost, unstructured and spontaneous science activities on the exhibition floors provide an effective forum for communication between the staff scientists and visitors. Media such as film, theatre presentations and school workshops offer a more structured method of communicating science. The centre receives over 200,000 visitors annually in a community of 150,000 people. This number rises to over 600,000 if Science North's outreach programme—which covers an area equivalent to twice the size of France—is included. Because of the close interaction between scientists and visitors, over 60 per cent of our visitors come to the centre more than once a year.

Science North has always taken an unconventional approach to science, not only in the method of presentation, but in its choice of topics. Controversial issues such as nuclear energy and nuclear waste, smoking and health have been tackled. We have collaborated with research scientists on environmental issues, particle physics, and technologies that assist people with sensory and physical handicaps.

The objective is to present visitors with coherent, understandable and objective information, accompanied by tools, demonstrations and computer programs that help them draw their own conclusions (figure 1). One example is our exhibition on nuclear fuel waste management. This is a hotly debated topic in Canada, particularly in Ontario which generates 90 per cent of Canada's nuclear waste. A deep underground waste repository is proposed to be built in the Canadian Shield bedrock of Northern Ontario. Science North worked with a variety of groups and individuals with different points of view. Representatives from Canada's nuclear agency (Atomic Energy Canada Limited) sat at the same table as members of Northern Ontario environmental groups as the exhibition content was discussed and refined. Although the exhibition did not take a stand on the issue, it provided visitors with facts and points of view that had

not been put by other media such as newspapers and television.

Neutrinos in everyday life

Unlike most fundamental particles, neutrinos were discovered thanks to their absence rather than their presence. The existence of the neutrino particle was first proposed in 1930 by the Swiss physicist Wolfgang Pauli as a way of reconciling data on beta radioactive decay with the laws of energy conservation, although the particle was actually named by Enrico Fermi.[3] Because neutrinos interact so feebly with other matter, it is extremely difficult to detect them and considerable ingenuity on the part of scientists and engineers is required to demonstrate beyond doubt that they have observed one in their detectors. The Observatory is one of the latest and most innovative of such detectors. Buried beneath 2 km of rock, shielded from virtually all extraneous radiation, the Observatory has the potential to answer some questions about the behaviour of neutrinos.

One of the most important issues is the solar neutrino problem. Experiments confirmed the existence of the neutrino in 1957. Since then, several detectors have been built to measure

Figure 1. Captain Neutrino: bringing the cosmos to everyday life

the flow of neutrinos from the Sun,[4] where they are created deep within its fiery core. However, these experiments have only been able to detect one-third of the neutrinos predicted by theory. In other words, there is a big discrepancy between what we think we know about neutrinos and what the experiments are indicating. This puzzling deficit of neutrinos challenges the validity of theories about the Sun and stellar evolution and, more critically, of our understanding of the structure of matter.[5]

The Sudbury Neutrino Observatory

In 1983, plans were put forward for an observatory at the Creighton Mine in Sudbury to study the feasibility of proton decay.[6] After funding was denied, researchers at the University of California at Irvine proposed exploiting the site to detect solar neutrinos using large quantities of heavy water. Thanks to the availability of the Creighton Mine, the loan of 1000 tonnes of heavy water from Atomic Energy Canada Limited, and collaboration between Canadian, American and European institutions, excavation began in 1988. The Observatory is designed to detect neutrinos by their interaction with the deuterium nucleus, which generates a minute flash of light.[7] More than 1000 photomultiplier tubes (PMTs) surround an acrylic vessel containing the heavy water, and can detect the light produced by the interaction.[8] The signals from the PMTs are then interpreted to determine the location of the interaction and its energy.

Data collection will start by October 1997. Currently, the Observatory scientists and engineers are working to assemble the acrylic vessel. Once that is completed, the remainder of the PMTs will be mounted. At this point the vessel will be filled with heavy water, and measurements will begin.

Collaboration

From the outset, Science North recognised the part it could play in providing public education about the Observatory. The two organisations set out to collaborate by:
- involving Science North's staff in the scientific and technological developments at the Observatory through regular contact with the Observatory engineers and scientists, and site visits.
- offering the public the unique opportunity to meet and participate in lectures given by leaders in neutrino research. Many such lectures have taken place over the past few years, presented by scientists such as Art McDonald, John Bahcall and David Schramm.
- developing exhibitions and presentations. With the help of joint fund raising, we have successfully developed a Can$30,000 (£14,000) exhibition on the experiment. We have also completed a Can$150,000 fund-raising campaign to develop a multimedia presentation about the Observatory.

Exhibitions and programmes

Science North is a mission-driven, consensus-based organisation. Our exhibit programme grows from the scientific interests of the staff, and major exhibition directions are discussed widely. For any major exhibition to become a reality, it has to have a champion who can argue its merits and develop support for the project among the science staff as a whole.

Neutrinos posed an interesting problem. On the one hand, the Observatory was clearly a major experiment that would focus considerable attention from the scientific community on Sudbury. However, the subject matter can be (and was) perceived by many to be dry, esoteric and inaccessible. This led to several important questions:
- Does the community care about the Observatory and about neutrinos?
- What would Science North have to gain by attempting to introduce visitors to neutrinos?
- How can a science centre explain the science of neutrinos?

The good news is that the answer to the first question is 'yes'. People in Sudbury are aware of the Observatory and are curious about its significance. There is a general understanding that the project is scientifically significant, and there is a certain level of community pride that it is happening here in Sudbury. Also, because the Observatory is being built at an operating mine, many Sudburians are close to the project or have neighbours who are involved. It is happening 'close to home'.

The answer to 'what would Science North have to gain?' becomes fairly straightforward in light of our mission statement which states that we will involve people in the science and technology of their everyday lives, especially as it relates to Northern Ontario. To ignore the Observatory would be to ignore our own mission statement.

Once the commitment was made to move ahead, the challenge posed by explaining the science of neutrinos required the minds of many to overcome. Staff worked closely with the Observatory scientists to develop an array of exhibits and audiovisual presentations which explore the world of the neutrino in an engaging, informative way. Exhibits were developed to illustrate the basic principles of the Observatory. These exhibits discuss the physics of neutrinos and the technology used in the Observatory, including detectors, computers and fine-particle measurement systems.

The exhibits

'Interpreting Heavy Water'. Heavy water contains an extra neutron in the hydrogen nuclei of each water molecule. It is the presence of this extra neutron that makes the detection and capture of neutrinos possible at the Observatory. Although it is very difficult to illustrate the reaction between the neutrino and heavy water in a hands-on manner, the difference between heavy and regular water is demonstrated by placing two containers on two digital scales. Both containers hold 1 litre of water. The visitor is challenged to guess which container has the heavy water. To confirm the visitor's guess, they can remove a

cover from the digital read-out of the scales and read the actual weight. One litre of heavy water is approximately 110 g heavier than regular water.

'Model and Computer Program'. A large model of the Observatory shows the construction and scale of the project. An accompanying interactive computer program was developed to provide design and construction information. The program offers some simple animations explaining the interaction between the neutrino and heavy water, as well as some of the technology used in the development of the project. The program also updates the status of the project. This section is complemented by a large bulletin board on which recent photographs, press releases and articles are posted.

'Photomultiplier Tube (PMT)'. This exhibit illustrates the function and the use of PMTs in the Observatory. A small light bulb inside a box produces light which is detected by a typical photoresistor and by a PMT. Both are connected to meters placed outside the box. An identical light bulb is placed outside the box and shines with the same brightness as the one inside. The visitor can vary the light intensity of these light bulbs with a control button. While changing the intensity, the visitor can observe the strength of the signal from the photoresistor and the PMT on the meters. The visitor notices that the PMT gives a very large signal when the light is extremely faint. The PMT is much more sensitive to light than the photoresistor or the human eye. Accompanying text explains the level of amplification produced by the PMT and the way it is used at the Observatory. This exhibit is complemented by an actual PMT, and visitors can see its construction. In addition, three other PMTs are suspended above the exhibit, mounted in a hexagonal honeycomb cell, just as they are at the Observatory.

'Amplification'. A simple hands-on exhibit explains avalanche processes in detectors and related phenomena. The exhibit consists of an adjustable ramp with horizontal edges placed at regular intervals along its surface. A digital scale is place at the end of the ramp. The visitor is challenged to arrange marbles on these edges so that the release of a single marble from the top will cause the maximum number of marbles to cascade down the ramp and be collected on the scale at the bottom. The arrangement of the marbles is critical to increase the probability of collisions. The inclination of the ramp defines the energy of these collisions which again has a role in the collision probability. The accompanying text explains the relevance of the activity to a PMT and other avalanche processes in physics.

'How Clean is the Observatory?'. The Observatory chamber has to be kept extremely clean (class 2–3: 30,000 particles per m^3) in order to eliminate radiation contamination. The contrast between this level of cleanliness and the mining environment surrounding the Observatory is striking and offers an ideal opportunity to discuss the technology used to maintain and monitor this cleanliness. The visitor uses a microscope to look at a collection of slides kept at different places for a month to collect dust. Sites included a woodworking shop, a typical living room, a classroom, a mine shaft, and the Observatory itself. It is easy to see the difference in the number of deposited dust particles on these slides. The same method is used by the Observatory to monitor their air quality inside the cavity. The accompanying text explains the importance of cleanliness for neutrino detection.

'The Observatory Object Theatre'. This 20-minute multimedia presentation on the Observatory is being developed. The show will be presented inside a mock-up of the acrylic

vessel, surrounded by PMTs. Objects, lighting effects, slides and video will tell the story of a particle physicist descending to the laboratory to conduct his research. The show also tells the story of a neutrino leaving the Sun on its way to Earth. The two eventually meet, when the scientist detects the neutrino in the Observatory. During the show, many of the Observatory's concepts and engineering successes, as well as its history, are touched upon. A video is being produced using the same general concept.

Conclusion

As I have emphasised, an exhibition on the Sudbury Neutrino Observatory at Science North is a matter of mission. The success, as with all our exhibitions, is proportional to the involvement and interest of our scientists in the project. It is ultimately the scientists that enrich the visitors' experience. Many of our staff have become very interested in the science, technology, progress and drama of the unfolding story of the Observatory. This interest and enthusiasm is key to the success of the exhibits. The curiosity of visitors is triggered by the exhibits, which typically leads to an in-depth conversation with our staff scientists. In formal settings, such as workshops for students, we have found that a greater level of understanding can be achieved through discussion. Students are generally fascinated by the challenges of building the Observatory, by the cosmological issues, by the researchers themselves and by the Observatory's impact on the community. In our experience, no exhibit can fully explain neutrinos, simply because of the fact that scientists themselves do not understand them. The only way we can involve visitors, is by giving them the opportunity to talk to scientists and use the exhibits as stimuli.

Notes and references

1 Carrigan, R A and Trower, W P (eds), *Particles and Forces: At the Heart of the Matter. Readings from Scientific American Magazine* (New York: WH Freeman and Co., 1990)

2 Shapin, S, 'Science in the making', *Public Understanding of Science*, 1 (1992), pp27–30

3 Pais, A, *Inward Bound: Of Matter and Forces in the Physical World*, (Oxford: Clarendon Press, 1986); Fowler, W A, 'What cooks with neutrinos?', *Nature*, 238 (1972), pp24–26

4 Folger, T, 'A question of gravitas', *Discover*, 16 (July 1995), pp38–39

5 Raghavan, R S, 'Solar neutrinos—from puzzle to paradox', *Science*, 267 (1995), pp45–50

6 Ewan, G T *et al.*, 'The Sudbury Neutrino Observatory—an introduction', *Physics in Canada* (1992), p112; Chen, H H, 'Electron detectors for the study of 8_B solar neutrinos', *AIP Conference Proceedings*, 126 (1985), pp249–76

7 Bahcall, J N, *Neutrino Astrophysics* (Cambridge: Cambridge University Press, 1989)

8 Hallman, E D and Evans, H C, 'The construction of the Observatory', *Physics in Canada* (1992), p130

Neutrinos in a science centre, part II— view from a scientist in the laboratory

David Wark

Neutrinos have long fascinated particle physicists. An experimenter at the Sudbury Neutrino Observatory near Science North, a leading Canadian science centre, explains why he is keen to share his fascination with visitors.

Why do some particles have mass, and others not? It sounds a simple question, but like many simple questions, it has proven frustratingly difficult to provide a simple answer. In fact, in the case of the neutrino it has even proven difficult to decide if the particles have a mass.

I am a member of a collaboration which is building an experiment, called the Sudbury Neutrino Observatory (SNO, figure 1), deep in a Canadian nickel mine to try to answer this question. We are also involved in a collaboration with Science North Centre in Sudbury, Ontario, to present the story of why we are building our experiment and how it is progressing directly to the public as it happens. We plan to display data from the experiment in Science North as they are taken. This article is intended to give the physicist's side of this collaboration. I shall begin by explaining why neutrinos are so fascinating to scientists working on SNO experiments. Then I shall give my personal views of why our collaboration with Science North is a valuable part of our work.

There are three known types of neutrino[1]—

Figure 1. Big science: the inside of the neutrino detector at the Sudbury Neutrino Observatory (see people, bottom right, for scale)

electron neutrinos, muon neutrinos and tau neutrinos—which are the partners of the electron and its heavier cousins the muon and the tau. Our current theoretical understanding of particle physics would indicate (for reasons we don't understand) that, of all the fundamental building blocks of matter, only these three types of neutrinos are massless. Physicists have therefore searched for any sign of neutrino mass since the day the particles were discovered, but so far have come up empty handed. As often happens in science, the route which we are trying to answer the mystery of the neutrino's mass was pioneered by a group which was looking for something else entirely.[2]

About 30 years ago, a group lead by the scientist Ray Davis began an experiment in a gold mine in South Dakota to measure the enormous neutrino flux produced by the nuclear reactions which power our Sun.[3] Normal matter is far more transparent to neutrinos than window glass is to sunlight, and these solar neutrinos can therefore travel directly out of the Sun, through space to the Earth, and penetrate deep under the ground to interact with Davis's detector.[4] It is the Sun's transparency to neutrinos that has provided the justification for the experiment. The energy produced by the nuclear reactions takes millions of years to leave the Sun, so when we look at the surface today we are seeing what happened in the core over the past 10 million years. Looking at the neutrinos, however, enables us to see what is happening inside the Sun right now.

The Davis group were very surprised when they observed fewer than a third of the number of solar neutrinos they expected. The first explanation was that something was wrong with the detector; however, years of testing have not revealed any instrumental reason for the shortfall. In fact, other solar neutrino experiments have now been done, one at Kamioka in Japan, a second at Baksan in Russia and a third at the Gran Sasso Laboratory in Italy. All have observed fewer solar neutrinos than predicted by models of the Sun. The second explanation, that these solar models are incorrect, is very hard to reconcile with the energy-dependence of the flux as measured by the four experiments. Another explanation exists, however, which is that electron neutrinos are emitted from the solar core in about the predicted number, but that a large fraction of them turn into other neutrinos (muon or tau) before they reach the Earth. Because the existing solar neutrino detectors have little or no sensitivity to these other flavours of neutrinos, this could explain why they measure too low a rate. This phenomenon, called neutrino oscillations, is interesting because it is forbidden in our current understanding of particle physics, and because it would happen only if neutrinos have mass. So if we could be sure we had observed neutrino oscillations, we would advance our understanding of these fundamental building blocks of nature and provide the long-sought confirmation that they are massive.

Neutrino oscillations provide an elegant explanation of the solar neutrino deficit, but how do we find out if it is the correct explanation? That is the purpose of the Observatory, which is being built by a collaboration of Canadian, US and British institutions.[5] An artist's impression of the detector is shown in figure 2. It consists of 1000 tonnes of heavy water (heavy water is D_2O, i.e. H_2O where the hydrogen is replaced by its heavier isotope, deuterium) contained within a thin acrylic sphere suspended in ordinary water. Within the ordinary water are about 10,000 photomultiplier tubes (light detectors sensitive to single photons) supported on a polygonal steel structure. To shield it from cosmic rays the whole structure sits in a huge cavity excavated 2 km underground in the INCO nickel mine near Sudbury. The idea is to detect two different nuclear reactions which solar neutrinos can produce when they hit the deuterons (a deuteron is the nucleus of a deuterium atom) in heavy water. The two reactions are:

1 neutrino + d→ p + p + e⁻
2 neutrino + d→ p + n + neutrino

where d is a deuteron, p is a proton, n is a neutron and e⁻ is an electron.

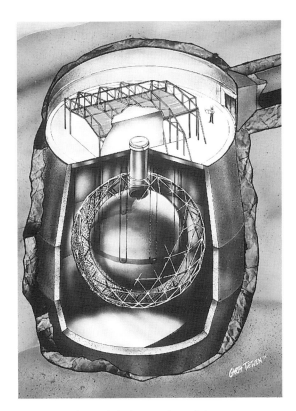

Figure 2. An artist's impression of the neutrino detector at the Sudbury Neutrino Observatory

The electron from the first of these reactions emits the optical equivalent of a sonic boom. This produces Cerenkov radiation, a faint flash of blue light which can be detected by the phototubes. By observing these flashes, we can count the electrons, and thereby count the neutrinos. The second of these reactions can be observed by detecting the neutron which is produced. The key to the experiment is that only electron neutrinos can cause the first reaction, while all types of neutrinos have equal probability of producing the second reaction. If only electron neutrinos are emitted by the Sun (as current theories of particle physics demand) then we should measure the same neutrino flux with both reactions. If, on the other hand, neutrino oscillations are the correct explanation of the solar neutrino deficit, we will see many more neutrinos with reaction 2 than with reaction 1.

This sounds simple enough, but in fact designing a detector which has good efficiency for detecting these relatively low-energy events, in the enormous mass of material necessary to get a decent counting rate while still keeping the background low (and measurable), has kept a team of about 75 scientists busy for more than 10 years.

Let me now try to answer a question which is directly relevant to our relationship with Science North: why do we want to do this experiment? I face that question all the time, and I think there are two good answers. The first is that a deeper understanding of the most fundamental aspects of the workings of nature is a good thing in itself. Some might say that if you could sway people by such an argument you wouldn't need to make it, but I think that people need to be taught to appreciate science even more than they need to be taught to appreciate art or music. That is the role of organisations like Science North, and anything we can do to help them is in our interest as well as everybody else's.

The second answer to the question is more practical. In the past, deeper understanding of the workings of nature has, in the end, been of direct benefit to mankind. I think of the money I put away for a pension as an investment in my future, I think of the money I will spend on my child's education as an investment in her future, and I think of the tax money which is spent on experiments such as mine as an investment in my great-grandchildren's future. Of course neutrinos are not the sort of thing for which one can easily see a practical use, so some might claim that I am being ridiculous, but such a claim ignores the interconnected and unpredictable nature of scientific advance. When Max Planck was doing measurements of black-body spectra, I am sure he didn't think he was playing an important part in the invention of lasers or digital computers—but those are just two of the things which can be traced by a long and tortuous path to his work. Just to give one example of this interconnectivity, consider that the same models of the Sun which our experiment is testing also predict that the luminosity of the Sun has considerably increased since life began on Earth. Scientists who are studying global warming often study the past climate of the Earth to gain clues as to how the climate may change in the future, and if they are operating under a mistaken belief about how the luminosity of the Sun has changed with time they may well reach the wrong conclusions. Thus one area of science affects others, and progress in one area can have unforeseen effects. It is the job of science museums to convey this insight to the public, and once again anything we can do to help we regard as time well spent. The taxpayers of Sudbury and Canada have spent many tens of millions of dollars on our experiment, without which we would be absolutely nowhere, so it

is really their science. Our project with Science North helps us get science out to the people who paid for it, and helps us to try to communicate to them the excitement and importance of contemporary fundamental science. Without such communication, the long-term support for projects like ours would be in danger.

I think the collaboration between Science North and the Observatory has been extremely successful, and I am certain that the overwhelming majority of scientists would be eager to be given an opportunity to have their projects presented in such a sympathetic and professional way by people who know the business of teaching science to the public. I hope that other science museums will consider following Science North's example.

Notes and references

1 Sutton, C, *Spaceship Neutrino* (Cambridge: Cambridge University Press, 1992)

2 Bahcall, J N, *Neutrino Astrophysics* (Cambridge: Cambridge University Press, 1989); Raghavan, R S, 'Solar neutrinos— from puzzle to paradox', *Science*, 267 (1995), p45; Bahcall, J N, 'The solar neutrino problem', *Scientific American*, 262 (May 1990), p26

3 In our current understanding of particle and solar physics, all of these neutrinos will be of the electron variety.

4 For the same reason, the overwhelming majority of those that hit this detector simply pass right through it, so even though the flux of neutrinos is quite large, scientists detect only a few a week.

5 Wark, D L, 'The search for solar neutrinos', *Endeavour*, 17 (1993), p104; Ewan, G T, *Nuclear Instruments and Methods*, 314A (1992), p373; Bahcall, J N, Calaprice, F, McDonald, A B and Totsuka, Y, 'Solar neutrino experiments: the next generation', *Physics Today* (July 1996), p30

Programming for success: people take centre stage

Melanie Quin

An exploration of unconventional approaches to bringing contemporary science to life through computer interactives, encounters with scientists and street theatre.

At the risk of gratuitous provocation, let me open with my definition of contemporary science, and my credo for promoting its understanding. Contemporary science is science which appears in the mass-media spotlight and so, however briefly, touches people's everyday lives. It is, by definition, framed by culture. Its context has social, economic, political and historical dimensions. I believe our message, when presenting contemporary science, should be a level removed from facts and figures. We should seek tools to draw the cultural framework, animate the debate and promote healthy scepticism over superstition and irrational thinking.

Techniquest, the leading science discovery centre located in Cardiff, has been experimenting with a vibrant and powerfully affective form of science communication: drama. Working as producer/director, dealing mostly with very practical problems, I have occasionally stepped back to apply the perspective of a broader science-centre experience. Are we developing anything others could build on? Practical tricks to hold an audience? A magical algorithm for the communication of contemporary science?

Reflection suggests four or five ways in which drama can be incorporated into exhibits or activities designed to touch people's lives. Be seated comfortably, and I'll paint you some pictures.

Software with a storyline

The traditional offerings of science centres and museums are real objects on display, and real processes with which visitors may interact: the magic of seeing a piece of moon rock at the Natural History Museum, London; the invitation to experience science and technology through several hundred hands-on exhibits in the cavernous Exploratorium, San Francisco. Over the past decade, science centres and museums have also developed increasingly sophisticated computer interactives. For museums that are prepared to take the risk of disturbing—even shocking—their public, the design of scene-setting environments and the development of engaging software now allows visitors to investigate context as well as scientific fact.

In 1992 at Heureka, the Finnish Science Centre, we developed a temporary exhibition for the 75th anniversary of Finnish independence.[1] *Finland 75 Years* invited visitors to take cover in a bomb shelter. As well as being given the opportunity to play the popular songs of the century on a juke box, they could find out what life was like living next door to Russia during the Cold War, and discover how Finland made the explosive break from agrarian past to post-industrial present within a span of two decades. The exhibition also addressed, head on, an issue both contemporary and historic: alcohol abuse. This theme was taken up in a computer game inviting visitors to compare their drinking habits with those of Finns through history, and with modern Europeans (figure 1). The exhibit, described below, proved to be one of the highlights of the exhibition, and has since found a permanent home at Heureka.

The game combines sophisticated graphics with a storyline spun from *Symposion*, a classic picture of four friends drinking at the Kämp restaurant, by nationalist artist Akseli Gallén-Kallela (1865–1931). An anonymous

Figure 1. A computer interactive developed by Heureka, the Finnish science centre, focusing on alcohol consumption

bourgeois has already passed out drunk, but Sibelius (the composer), Kajanus (the conductor) and Gallén-Kallela himself are still in animated conversation. Interaction is simple: point the mouse and click. If you click on the figure of Gallén-Kallela you will discover the social scandal that surrounded the public showing of the picture. 'It was a grenade that exploded, and all the shrapnel hit me,' wrote the painter, 'and yet I exposed my inner self.' Click on the sleeping drunkard and the dangers of alcohol abuse are portrayed: from impotence and cirrhosis of the liver, to traffic accidents and social problems. Statistical charts, and the 1920s poster campaign of the Milk Propaganda Office, underline the message. Choose Kajanus, and several apocryphal stories are revealed. One legend has it that Kajanus left the table to conduct a concert in St Petersburg. Returning next day to find the company still assembled, he was scolded by Gallén-Kallela who told him 'Don't you jump up and down all the time, be seated!' Click on Sibelius and his image on the 100 Finnmark banknote appears. Today the tax on alcohol accounts for some 10 per cent of state revenue. In the eighteenth century the consumption of foreign wines was declared 'dangerous' because it was draining the country's gold reserves. Prohibition, enforced from 1919 to 1932, resulted in a tenfold increase in the consumption of medicinal alcohol, and backfired as the law fell into disrepute, the bootleggers grew rich and millions of litres of absolute alcohol flowed in from Poland and Estonia.

But what of drinking habits today? Click on the restaurant table, and you'll be offered unlimited beer, wine or the Finnish favourite— gin and bitter lemon. Drink as much as you like for one hypothetical week: how much weight will you put on if you keep drinking like that for a year? And how does your consumption compare with that of French wine lovers or German Bierstube regulars?

Such an approach to exhibit development is a lot less unconventional to non-English speakers. The Finnish word for 'science' is 'tiede'. The word spans the full spectrum of human knowledge from mathematics to humanities, and so do the programmes at Heureka. The word is derived from 'tietää' meaning 'to find the way', 'to get to know'. The Dutch 'wetenschap' and German 'Wissenschaft' are similarly broad. Colloquial English is hampered by a narrow natural-sciences interpretation of the word 'science', as underlined by reactions to the exhibition *Science in American Life*, at the Smithsonian's National Museum of American History.[2]

Computer interactives inviting role play

Three-dimensional exhibitions, aided and abetted by engaging computer interactives, are all very well as far as they go, but in combination with live interpreters we go further.

At newMetropolis, the Netherlands national science and technology centre,[3] software developer Martijn Plak and I have begun work together on 'Synthesis', a computer game designed to involve visitors to a 1200 m² *Energy* exhibition in an energy-management challenge.

The game was to place the visitor in the role of a national government which must match the country's projected energy consumption with adequate supplies, over several decades. To do so, the economic, social and environmental costs/benefits of the national energy-management strategy would have to be examined and balanced. The international implications of the national strategy would also have to be weighed. At the game's conclusion, the resulting energy world would be evaluated: how green the land, how clean the water and air, how prosperous and contented the population?

The experience of a year of prototyping and budgeting led Martijn Plak to revise his thinking on whether it is possible and cost effective to present complex ideas on contemporary science through meaningful, interactive exhibits. He concludes that there are good interactive exhibits but, at the price, people are better:

Presently, science centres present science as facts, accept those facts as THE TRUTH, and offer the visitor a wonderful experience discovering those facts. What they should do is take the message they convey a level up, and make science visible . . . question its character . . . and help us understand its position in our lives. I think programmed activities are the way to go: workshops and presentations with a close contact between visitors and Explainers. Anything can be explained by a good story-teller, and more important than the software developer is the job description of the Explainer. He or she is like an old sailor who tells you about far-away places in the time they were really far away. There is a small inconsistency: a story takes you somewhere. Here the story is changed according to the responses in the audience. The Explainer must pick up subtle responses, and get people to react.[4]

In other words: no explainer, no opportunity for open-ended dialogue.

Face-to-face encounters with scientists

If our visitors are citizens with whom we wish to engage in dialogue, we need to make eye contact. And without question, visitors appreciate face-to-face encounters. 'The best exhibit is a real object and a real person,' says David Taylor, exhibits director at the Pacific Science Center, Seattle.[5] A genuine enthusiast is, of course, irresistible. Jim Marchbank, executive director of Science North, Sudbury, concurs: 'The more people–people interaction we offer, the happier the visitors.'[6]

These observations suggest a logical extension to the 'peopling' of science centres: invite scientists to speak on contemporary issues, not only to hushed audiences in a lecture hall but also to family groups on the exhibition floor. Demonstration areas offer perfect sites for informal encounters and the questioning of 'real' scientists.

Dramatic encounters

The use of drama and human interpreters at the Science Museum, London, suggests that this 'low-tech medium' might just be the most effective way of ensuring that visits to science exhibitions actually stir the emotions and thus become memorable.[7] Conversations overheard during the interval of a performance of Tom

Mouldy Mind-Reading Sketch

Music Theme from The X Files

Chris (at lectern close to microphone) And now we enter the spooky world of the paranormal. Of communication through extrasensory perception. You know the kind of thing. Dodgy Russian psycho-kinetic professors and cheesy American mentalists, mouldy old mind-reading acts and terrible telepathy, fake fakirs, and stupefying spoon benders, hyped-up hypnotists and . . . well I think you all get the idea.

Enter Gary and Fran stage left. Gary sets chair centre stage—Fran sits. She wears a long dark dress, colourful shawl and carries a blindfold. She has the air of a seaside palm reader.

Fran (dramatic voice) I need a volunteer please.

Assistant (Gary) goes into audience to find a volunteer.

Fran I will now put on the blindfold *(she does this)*. Ask the volunteer to choose a personal item. Anything they have about their person, in their pocket or handbag.

Gary Their what?

Fran (as Edith Evans in The Importance of Being Earnest*)* Their handbag!

Gary (to people in the audience, rejecting anything other than a pair of spectacles or a watch) Haven't you got anything else?

Fran (impatiently) What's going on? Do hurry up!

Gary borrows a watch from someone and gets them to hold it up.

Gary There is not much time left for you to guess this item.

Fran Oh . . . mmm . . . um . . . I smell something—yes some herbs. Is it some thyme?

Gary (pulling a face) Not quite. Now is the hour for you to discover what I am holding in my hand. It's not a brand new item. It may be second hand. It has a second-hand feel to it.

Fran Ah yes, of course. A second-hand heel. It's a pair of shoes *(chuckling)* perhaps not bought too recently *(she laughs)*.

Gary Ladies and gentlemen, the Great Madame Sillyname doesn't always have the muse with her. So let's give her one more chance.

He uses a second volunteer to hold up a pair of spectacles—these are the only items he can 'transmit' and he is none too confident of Fran's ability to get even these two right. Gary is prepared to produce his own spectacles from a pocket and to encourage the volunteer to hold these up.

Fran Is there another item?

Gary There is, but be careful with these, people may see through what you are doing. They are quite a glassy pair, I mean a classy pair.

Chris is in the audience. He holds up 'Boo!' and 'Hiss!' signs. The audience responds by booing and hissing.

Gary I think it's time to leave.

Fran Leaves . . . pages of a book?

Gary No . . . to leave . . . go!

Fran gets the idea and they both scuttle off stage left.

Stoppard's play *Arcadia* revealed to me just how powerfully a top-flight playwright can stimulate a West End audience to scientific debate.[8]

Yet both museum and theatre present hurdles of accessibility. If you want to go to the theatre, you have physically to go there. Although less intimidating than a research institution, museums can still be perceived as 'temples' of science.

With the twin aims of taking science to where people are, and bridging the art–science cultural divide, the PanTecnicon project was conceived. PanTecnicon is a celebration of 'the Arts applied in Wales', and will build to an annual season of popular art–science events throughout the principality. Summer 1996 was the pilot season and performance was the first medium we explored, with a touring show that owed more to television culture and street theatre than to Tom Stoppard. One sketch from the script is shown opposite.

The theme of the show was communication: language, gesture, Aldis lamps, semaphore flags, mime, melodrama and, as illustrated in the sketch, extrasensory perception—which can certainly be tagged with the 'contemporary science' label. A poll reported in *Nature* exposed the fact that 48 per cent of Americans believe in unidentified flying objects, and 29 per cent are convinced that contact has been made with aliens.[9] In the UK, around half the population studies horoscopes, and astrology columns are an accepted component of magazines and newspapers.

The way we choose to live, what we value, and what we don't, are decisions informed by more than rationality, as Brian Eno (commenting on his speech at sculptor Damien Hirst's Turner Prize award ceremony) fluently argued:

We somehow have to arrive at decisions about those things that can't be separated out from their context, things of which we don't even know the boundaries, things that are vague, complicated and mostly unknowable. How, for instance, does one arrive at a feeling about, say vegetarianism? Well,

partly through 'rational discussion', but mostly through a complex bundle of stylistic choices—through taste. And taste is evolved as much by soap operas and Damien Hirst's split cows and BSE scares as it is by rationality.[10]

As artists, dramatists and science communicators, the PanTecnicon team believes science to be the disciplined search for the truth, and prizes rational thinking. Hence a script using humour to highlight credulity and debunk nonsense. Hence also the choice of street theatre as the medium. Visual and physical, circus and serious: it's a vibrant medium in which eye contact and sharing with the audience are paramount. Participation elements and games amuse, inform and—significantly—unite the audience: this is a prime consideration and also the reason why the audience is frequently addressed directly. This is not television, it is really happening. The tensions are real, and the performers work with that bond.[11]

Conclusions

I have argued in the past that flashy exhibits alone cannot, and should not be expected to, capture and hold the interest of twenty-first-century audiences, and that the design of spaces for social interaction and pathways for individual exploration, in galleries peopled with friendly explainers and real scientists are the prerequisites for a meaningful museum experience.[12]

Today, I would go further, taking the action beyond the museum walls. To communicate contemporary science effectively, we must recognise that the exhibition medium is limited by the lead time required for development, and that it is in direct competition with television programmes (produced much faster) and World Wide Web sites (updated much more easily). We must also recognise that the museum is only one of the places available to engage our public.

Viewing science as part of culture and the public as citizens with whom to enter into dialogue, we should no longer evade contentious issues but, rather, be prepared to give up objectivity and admit to having a viewpoint. We must also exploit the communication media best suited to such meeting places and such forms of dialogue. Theatre in the streets—performed by player–scientists, designed to spark and animate debate—offers an enticing medium to explore.

Notes and references

1 Heureka's mission is 'to stimulate and promote interest in science and research, by providing impressive visitor experiences in an attractive, safe environment characterised by high-class and friendly service'. The science centre also aims to 'surpass visitors' expectations, and to encourage them to come again'.

2 Molella, A, this volume, pp131–37; Molella, A, and Stephens, C, 'Science and its stakeholders: the making of "Science in American Life" ', in Pearce, S (ed), *Exploring Science in Museums* (London: Athlone, 1996), pp95–106

3 For more information on newMetropolis see Bandelli, A and Bradburne, J, this volume, pp181–87.

4 Plak, M, National Science and Technology Centre, Amsterdam, the Netherlands, personal communication (1996)

5 Taylor, D, Pacific Science Center, Seattle, US, personal communication (1996)

6 Marchbank, J, Science North, Sudbury, Canada, personal communication (1996)

7 Farmelo, G, 'Drama on the galleries', in Durant, J (ed), *Museums and the Public Understanding of Science* (London: Science Museum, 1992), pp45–49

8 Stoppard, T, *Arcadia* (London: Faber and Faber, 1993)

9 Nadis, S, 'Sceptics course seeks to stem "tide of irrationalism" ', *Nature*, 382 (1996), p199

10 Eno, B, 'Getting the picture', *W Magazine* (London: Waterstones, 1996), pp12–20

11 Fox, S, 'Street and outdoor performance', in Coult, T and Kershaw, B (eds), *Engineers of the Imagination* (London: Methuen Drama, 1990), pp31–41; Pledger, C, *Signal, Mirror, Manoeuvre: A Celebration of Communication* (Cardiff: Techniquest, 1996)

12 Quin, M, 'Editorial', *ECSITE Newsletter*, 21 (1994), p2

Lessons of *Science Box*

Lorraine Ward

At the Science Museum, London, several topics in contemporary science and technology have been tackled in two series of small modular exhibitions. What lessons have been learnt?

'It sounds as dull as ditchwater, actually.' The Science Museum's press officer was not impressed with the exhibition plans devised by the 'Chaos' exhibition team. We were outraged—'What does she know about it, we're the science professionals.' One week later, the team conceded that the press office had a point, and we embarked on making 'Chaos' more accessible to the general family visitor. It had become clear that, unless we incorporated the public viewpoint by displaying material such as the UK's new National Lottery number-selecting machine, visitors would come away with the view that the science of chaos had little relevance to everyday life except to keep mathematicians off the streets. The lesson here was that it is important never to lose sight of the public's view of science if an exhibition is to communicate its central message.

This paper will concentrate on discussing the series of contemporary science exhibitions at the Science Museum, which are aimed at a general family audience and marketed under the brand name of *Science Box*. It will look at the issues that the series has faced: legitimisation, exhibitability and outreach beyond the confines of the Museum. From this, conclusions will be drawn which will be relevant to future exhibitions on contemporary science in the Science Museum and also for others working in this area.

Two major factors work against the representation of the public viewpoint in the Science Museum. The first factor is the role of the Museum within the scientific establishment, and the second is the issue of what constitutes exhibitability. The sentiment that 'We are the Science Museum, after all' has been a restrictive factor in what we do.[1]

The Science Museum, as a national institution, is expected to provide neutral, authoritative views on science.[2] This is one of the reasons why sponsors wish to be associated with it. The obligations and public expectations of establishment institutions, however, make it more difficult for museums to put science in its social context and to deconstruct science and scientists.[3] Fulfilling obligations raises other questions. How far should an exhibition go to satisfy a sponsor? Should it feature an obscure piece of uninspiring research equipment to win the endorsement of the scientific cognoscenti? How much notice should the team take of a scientific advisor's fundamental disagreement with the angle and direction of an exhibition?[4]

The second factor of exhibitability—providing a strong visual and intellectual impact for the visitor—presents other challenges. In terms of visual impact, the visitors' expectation of being provided with a new type of experience is confounded when contemporary objects come in the form of uninspiring packages resembling shoe boxes or coffee percolators. The more exciting possibilities, such as sculptures and interactives, are often constrained by the limitations of time and money. In terms of intellectual impact, much revolves around the debate about what constitutes scientific literacy. Should the exhibition be didactic or issue-based, and can the topic be presented without incorporating many basic scientific facts? Are facts always central to the understanding of scientific issues? Should an exhibition say something about scientific methodology—and, if so, would a public seeking 'infotainment' really be inspired by an exhibition that states that scientists spend rather a lot of time publishing papers, seeking grants and disagreeing with each other?

The origins of Science Box

If a visitor wanted to explore a contemporary science topic, he or she would be unlikely to find it at the Science Museum. In this shrine to technological achievement, computing stopped in 1981, on the eve of the break-through of the IBM personal computer. Galleries such as *Computing Then and Now* come to a halt in the year of their completion, with further renewal dependent upon additional sponsorship. Even when update areas are incorporated into a gallery, a permanent exhibition will inevitably date from the day of its completion.

Until the end of 1991, the Science Museum did not have a mechanism for addressing contemporary, newsworthy science topics. The Museum organises much of its work for the public under the terms of our 'Public Understanding of Science' initiative. As part of this initiative, Professor John Durant (Assistant Director and Head of Science Communication) was asked to set up a Museum-wide project team. The team was given the task of designing a modular display system for use in the production of small exhibitions on contemporary science. The original concept was that *Science Box* exhibitions should be 'jewels': small displays encapsulating the latest in scientific research. The subject matter would be topical, with the exhibition being produced within weeks of a science news story breaking.[5] Sending an exhibition on regional tours across the UK was an opportunistic afterthought.

The speed at which the project was realised was fortuitous for the Science Museum. Within weeks of the concept and sponsorship proposal being drawn up, the idea was taken up by Nuclear Electric plc, who sponsored a series of seven exhibitions over two years. The arrangement with Nuclear Electric was later extended for a further two years. Additional sponsorship—whether for touring or for the on-site exhibition—became a prominent feature of the second series, leading to the generation of larger and more complex exhibitions such as 'Big Bang' and 'Information Superhighway'.

What is Science Box?

The name *Science Box* is reflected in the distinctive modular design. Each exhibition within the series (table 1) is constructed around a reusable carbon-fibre 'box' framework covering an area of up to 50 m², which can be dismantled and reconstructed in a new shape for the next exhibition. The framework is designed to carry lighting, text and graphics, and to house computers, electromechanical interactives and objects. The average *Science Box* exhibition comprises 15 text panels of 70 words each, a 1000-word leaflet, two computer touch-screen interactives, one video, two hands-on interactives and two objects.

The aims of *Science Box* were:
• to provide a series of small, rapid response exhibitions that give the non-specialist public an introduction to contemporary issues and ideas in science, medicine and technology
• to provide an explanation of scientific topics currently featured in the media
• to illustrate the importance of such topics to the everyday world of the visitor

Legitimate science?

There are problems associated with accepting sponsorship for science and technology exhibitions. When an organisation sponsors an art exhibition, it is understood that the company is raising its profile and making use of public-relations opportunities by providing funds for the exhibition to be realised. In science and technology exhibitions, the sponsor often has a direct interest in the subject matter and may expect to influence the content. Sponsorship contracts at the Science Museum state that final decisions on the content and editorial control lie with the Museum, but sponsors are also regarded as partners and therefore expect to have a role in development.

The relationship with Nuclear Electric plc was successful because the company's main aim was to improve its public relations through its association with contemporary science. The absence of sponsor pressure over topic selection enabled the exhibition planners to create an overall balance between the physical and life sciences, and allowed scope to respond to new opportunities. Other factors influencing the choice of topics came into play, for example the Government Office of Science and Technology offered to fund a tour of 'Information Superhighway' because it dealt with information technology, and 'Passive Smoking' (figure 1) promoted the Science Museum's no-smoking policy. However, the sponsors were not entirely disinterested in the topics of the exhibitions. For example, Nuclear Electric's public-relations advisors expressed concern over the suitability of 'Alzheimer's Disease' because it was about elderly people and was therefore 'depressing and unappealing to a younger audience'.

The sponsor's fears were successfully allayed and the exhibition went ahead.

Sponsors are attracted by the good public relations that accompany being identified with dynamic and progressive science, and/or because their products or fields of interest are reflected in the exhibition content. The *Science Box* team holds consultations and 'brainstorming' sessions with scientific experts to ensure that the scientific community broadly approves of what we do, to give our exhibitions a stamp of authority and to confirm technical accuracy. In our experience, although this might produce a false perception of expert consensus,[6] highlighting scientific disagreement does not necessarily assist the public's understanding of the topic. This does not mean that the Museum has been afraid to confront controversial issues. Problems with technologies are discussed, but not at the expense of portraying science and technology as positive, progressive, dynamic and exciting. For example, 'Passive Smoking' had all the ingredients for success—it was topical and

Table 1. *Science Box* exhibition topics and duration

Series 1

DNA Fingerprinting, 1 March to 29 June 1992
Living with Lasers (applications of modern laser technology), 15 July to 30 September 1992
Ozone—a Cover Story (thinning of the ozone layer), 13 October 1992 to 10 January 1993
Passive Smoking, 14 January to 15 March 1993
Superbike (the UK's Olympic-gold-medal winning bicycle), 16 March to 20 June 1993
How Small Can We Go? (nanotechnology), 29 June to 26 September 1993
Speak to Me (technological aids for people with disabilities), 4 October 1993 to 30 January 1994

Series 2

City Limits (sustainable cities), 24 May to 2 October 1994
Infertility Maze (infertility and its treatment), 19 October 1994 to 26 February 1995
Information Superhighway (global communication technology), 26 April to 3 September 1995
Chaos (the science and applications of chaos theory), 3 November 1995 to 14 April 1996
Alzheimer's Disease, 2 May to 1 September 1996
Big Bang—Birth of the Universe, 5 September 1996 to 5 January 1997

Figure 1. The 'Passive Smoking' Science Box *exhibition at the Science Museum, London*

controversial, it had a heated opening debate, and it attracted visitor interest and press appeal. In this exhibition, the team came out firmly on the side of the anti-smoking lobby, raising concern over whether the Museum should present an opinion, rather than a consensus, to the general public.[7] The team's view was that the presentation of an opinion was a positive move for the Museum in that it found favour with visitors and it generated publicity for both the Museum and the exhibition topic.

Exhibitable science

Visitors, on average, spend less than two minutes in a *Science Box* exhibition. Visitor studies—quantitative and qualitative surveys of visitor patterns and behaviour—were incorporated into *Science Box* development. Most of the early exhibitions had full summative evaluations of the completed exhibition which

revealed to what extent they had met their aims. Evaluators examined the messages understood by visitors, mapped time spent in the exhibition and visitor flow, and identified 'hot spots' to find out which were the most-visited components.[8] From these studies, later teams were able to see what lessons were learnt and what they themselves could apply to subsequent exhibitions. For example, visitors' views on interactives (most popular) and on text (least popular) were important to bear in mind for the development of future exhibitions.[9] The two-minute staying time was indicative of success, because this was long enough for visitors to digest the main message of the exhibition, and it compares well with the average of five minutes spent by a visitor in a major permanent gallery at the Science Museum.

Awareness of visitors' viewpoints is important in the wider context of the debate on what constitutes scientific literacy. How contemporary is the selected topic? Should we select topics from the standpoint of the current scientific agenda, or from the standpoint of public interest? If we set an agenda, would there be an audience for it? Should we aim for consensus? Should we ignore controversies or should we highlight them? If we choose to focus on scientific process, would the public really be interested? If we intend to bring out social and ethical issues, how do we do so without covering a vast exhibition space with textbook science? Can we be promoting scientific literacy whilst avoiding a didactic approach?[10]

With so many competing criteria and so little space to incorporate them, the waters are bound to get muddied. 'But we are the Science Museum, after all'[11] is the cry, as the space for issues is sacrificed to the didactic explanation of the mechanics of basic science or technology.[12] The dilemma is that it is felt that the public cannot judge issues without knowing the facts behind them, and visitor surveys demonstrate that the public say that they would come to an exhibition to get the 'facts' behind the news (although 'facts' are generally elusive in science).[13] Conversely, these same surveys also show that the traditional 'how-it-works' approach arouses the least interest with visitors, because visitors are more concerned with how science will impact on their own lives.

The criteria for exhibition content used by the team are intellectual: each text panel, interactive or object must have a reason to be there, yet at the same time the content has to be developed with the visitors' staying power in mind. A variety of interpretative media makes a livelier and more appealing exhibition, both visually and intellectually.[14] This also presents practical challenges in fast-turn-over exhibitions, because of the timescale and cost that it takes to produce such exhibits. Yet this view of exhibition content is still perceived as being 'radical' to some parts of the museum community.[15] After a *Science Box* opening, a colleague would comment to the team, 'It's not

a proper exhibition, though, is it? I mean, where are your objects?' Exhibitable, meaningful, inspiring objects central to the main message of an exhibition are thin on the ground in the field of contemporary science, but because 'We are the Science Museum, after all' there is an expectation by colleagues—not necessarily by visitors—to incorporate objects.[16] Whether the term 'objects' also embraces living or dead human tissue is another matter for debate. A team meeting on the 'Infertility Maze' exhibition was spent discussing the ethics and problems of using live sperm in conjunction with a sperm-tracking machine. The problems of sourcing and maintaining such material led to us taking the easier route of using a video instead. The visitor evaluation of the exhibition demonstrated that the most popular item in the exhibition was the 'Sperminator'—a bagatelle game designed to convey the probability of conceiving normally (figure 2). Simple hands-on interactives such as this attract both children and adults—if an exhibit does not appeal to children, even if the adults are interested, the group will not stop during a family visit. This is a reflection of the tension within the museum community between the museum as a celebration of material culture, or as an interactive science centre.[17]

In the light of experience, the *Science Box* team established a formula for contemporary science exhibition production. In practice, this comprised a portfolio of designers, interactives producers and other consultants. Also, a set of interpretative rules was gleaned from successive evaluations, for example, no panel text should be longer than 70 words. A formula guarantees efficient and rapid production of exhibitions to a high standard, but it has many drawbacks. It can restrict creativity in terms of both design and content, and can lead to stagnation.

Outreach

The development of any *Science Box* exhibition involves liaison with the Museum's press office. One of the main aims of *Science Box* is

Figure 2. The 'Sperminator' bagatelle game in the 'Infertility Maze' Science Box exhibition

to generate maximum publicity for the Museum and its sponsors. A carefully planned press strategy can sometimes generate at least as much publicity for a *Science Box* launch as for the opening of a major gallery. The launch strategy is decided in the early stages of exhibition development, and usually consists of a press-preview breakfast with speeches by a celebrity and/or an eminent scientific advisor. For example, the 'Alzheimer's Disease' exhibition opening featured the respected clinician Dr Martin Rossor, and actress Britt Ekland who spoke emotively of her mother's suffering from the disease. An effective press strategy does bring issues in contemporary science—and the Museum's role in promoting them—to a wider audience. However, there is a drawback: the more effective the press strategy, the greater the number of complaints from visitors who expect to see an entire gallery devoted to

their subject of interest. The production of publicity and associated print material for the exhibitions does present other problems. Sponsors expect to see a large amount of print material bearing their logo, and in many ways they regard the print content as being of greater significance than the exhibition content.

In 1991, the Science Museum did not envisage that its flagship touring exhibitions would be vomited over in shopping centres. Although the *Science Box* structure was designed to tour, the touring programme itself emerged haphazardly. The withdrawal of the long-standing annual government grant to the Museums and Galleries Commission's travelling exhibitions service was substituted by a one-off payment to fund the tour of 'DNA Fingerprinting'. The tour of 'Passive Smoking', sponsored by the Department of Health, gave touring a new direction. Whereas 'DNA Fingerprinting' visited mainly large regional museums and science centres, the brief from the Department of Health to reach new, wider audiences meant that 'Passive Smoking' toured shopping centres, hospitals and community centres.

Designing an exhibition to tour does place restrictions on the design of the original exhibition: a complicated design, exhibits which require frequent maintenance, and more valuable or fragile material are not suitable for touring. The three days of set-up time by a team of three technicians, plus one day of take-down time and travelling time, mean that short-stay venues of less than one month are not usually viable.

Touring was evidently a way of bringing in extra sponsorship, but it also became apparent that sponsors preferred the relatively cheaper option of sponsoring a tour of an old exhibition, rather than paying for the development of a new exhibition. Yet the price quoted to sponsors is relatively inexpensive only because touring depends on using the Museum's own established infrastructure. Even after staff time has been fully costed, it is less than a third of the cost of contracting out to a facilities house.

This means that, at present, no more than two exhibitions can be on tour at any one time, and we cannot meet the existing demand for sponsored tours.

Lessons learnt

'We should be more like the BBC, not Sky TV,' proclaimed a colleague on the future direction of *Science Box*. In other words, *Science Box* should be taking risks, presenting science and technology which is 'bubbling under' to the general public. How can we do this without becoming didactic and without falling into the eternal trap of being seen as legitimising science and scientists? If an exhibition—or indeed, any other media—is to communicate effectively, then it must be aware of visitor interests, preoccupations and prejudices, whatever they may be. It would be unwise to lose sight of the fact that many visitors regard the paranormal and science fiction as legitimate topics which they would expect to see in the Science Museum. Visitors come to enjoy themselves, and exhibitions therefore need a sense of humour. Infertility was a subject which could have been earnest, worthy, serious and dull. For a topic which hardly lends itself well to interactivity, the most popular and effective exhibit in the 'Infertility Maze' exhibition was the 'Sperminator' bagatelle game. Such interactivity does not necessarily conflict with the Science Museum's role in the scientific establishment, but does put exhibitors in the spotlight to communicate, to the establishment and to sponsors, the rationale for this approach.

Security of long-term funding and continuous striving to inform sponsors and generate an atmosphere of trust is essential to the success of an exhibition series. Sponsor interest in touring has made us look again at the logistics and strategy of our touring programme. The series of exhibitions now being produced as successors to the *Science Box* series will have a very different relationship with sponsors. There was strong evidence to suggest a demand by individual sponsors for individual *Science-Box*-style exhibitions on specific topics, with touring as an add-on extra. This is the new strategy being pursued, with exhibition slots being booked by sponsors in advance. The challenge that the Museum now faces is that each sponsor has their own agenda in science communication, and each will want to play a part in the process of devising the intellectual framework of the exhibitions.

There is still much to be learnt. On the intellectual side, tackling issues and broaching topics such as scientific methodology are to be important in the new series. We still do not respond quickly enough to news items, and we need to be always alert to the dangers of stagnation. On the practical side, we need to generate fresh design approaches, build interactives more quickly and cheaply, and tour our exhibitions more effectively. The new series will present all these challenges, and many more.

Notes and references

1   Macdonald, S and Silverstone, R, 'Science on display: the representation of scientific controversy in museum exhibitions', *Public Understanding of Science*, 1 (1992), p82

2   For a review of museum orthodoxy, see Finn, B S, 'The Museum of Science and Technology', in Shapiro, M S, *The Museum: A Reference Guide* (New York: Greenwood Press, 1990).

3   See Holton, G, *Science and Anti-Science* (Cambridge, MA: Harvard University Press, 1994), for a view of science and its critics.

4   Scientists are often very conservative in their approach to the public and the

media: see van den Brul, C, 'Perceptions of science: how scientists and others view the media reporting of science', *Studies in Science Education*, 25 (1995), pp211–37.

5 A Royal Society report suggested that exhibitions could possibly be used for presenting issues of current concern, thereby contributing to more scientifically informed discussion: see Bodmer, W, *The Public Understanding of Science* (London: The Royal Society, 1985), pp27–28.

6 Macdonald, S and Silverstone, R

7 Ross, M, 'Passive smoking: controversy at the Science Museum', *Science as Culture*, 5 (1995), pp147–51

8 Ten *Science Box* exhibitions, held in the Science Museum, London, were summatively evaluated; the reports are unpublished.

9 For the long-term impact of interactives on visitors, see Stevenson, J, 'The long-term impact of interactive exhibits', *International Journal of Science Education*, 13 (1991), pp521–31.

10 This is part of a much wider debate about what constitutes scientific literacy. For the orthodox view, see Miller, J D, 'Scientific literacy: a conceptual and empirical view', *Adedalus*, 112 (1993), pp29–48. For a counter-orthodox view, see Shamos, M, *The Myth of Scientific Literacy* (New Brunswick: Rutgers University Press, 1995).

11 Macdonald, S and Silverstone, R, p82

12 'Most science museums devote energy and resources to communicating the facts and concepts of science and technology through the most up-to-date and imaginative interpretative and aesthetic methods,

without so much as a nod in passing to the ethical problems the subject matter raises': Carnes, A, 'Showplace, playground or forum?', *Museum News*, 64 (1986), p30

13 Macdonald, S and Silverstone, R, p84

14 Mintz, A, 'That's edutainment', *Museum News*, 73 (1994), p33

15 Multimedia is no remedy for an uncreative approach: see Kavanagh, G, 'Dreams and nightmares: science museum provision in Britain', in Durant, J (ed), *Museums and the Public Understanding of Science* (London: Science Museum, 1992), pp81–83.

16 The value and role of objects in exhibitions has attracted great debate. For example, Silverstone sees them as having a 'magic potency', whereas Lowenthal's view is that 'objects explain little about the genesis of scientific ideas, processes or technologies they illustrate . . . and seldom address the economic and social impacts of science and technology.' Silverstone, R, 'The medium is the museum: on objects and logics in times and spaces', in Durant, J (ed), *Museums and the Public Understanding of Science* , pp35–36; Lowenthal, D, 'Science museum collecting', in *Museum Collecting Policies* (London: Science Museum, 1991), pp11–12

17 For a reference source on interactivity and its virtues, see Gregory, R L, 'Turning minds on to science by hands-on exploration: the nature and potential of the hands-on medium', *Sharing Science: Issues in the Development of Interactive Science Centres* (London: The Nuffield Foundation on behalf of COPUS), pp1–9.

The Lemelson Center for the Study of Invention and Innovation: programming in action

Arthur Molella

The Lemelson Center runs creative outreach programmes that meet the challenge of placing invention and innovation in multiple sociocultural contexts without losing sight of scholarship.

This paper describes the Smithsonian's new Jerome and Dorothy Lemelson Center for the Study of Invention and Innovation. The Lemelson Center is a programme within the National Museum of American History, itself one of a dozen museums under the general umbrella of the Smithsonian Institution. I will argue that presenting current science and technology in historical and sociocultural context is a highly effective way to reach broad audiences and that museums are, therefore, among the most effective vehicles for doing so. Here, I particularly distinguish science museum expositions from the sorts of decontextualised demonstrations favoured by science centres. In making this argument, I am fully aware of the increasing and welcome encroachment of science-centre-style programmes into museums—including both the Smithsonian and the Science Museum, London. And, as an employee of the Smithsonian during the great public row over the exhibit of the Enola Gay, the B-29 that dropped the atomic bomb on Hiroshima, I am also painfully cognisant of the dangers of historical interpretation in public museums.[1] On balance, though, for achieving the goals of public understanding, I believe that the benefits of contextual presentation far outweigh the risks.

The Lemelson Center, which opened in January 1995, is a benefaction of wealthy independent inventor, Jerome Lemelson, holder of well over 500 patents relating to robotics, machine vision, the video-cassette recorder and other electronic wonders, qualifying him as one of the most prolific inventors in American history.[2] Deeply concerned that the US was losing its technological edge in the global economic competition, Lemelson launched a national educational campaign a few years ago to reverse the trend. In particular, he approached the Smithsonian to tap its considerable outreach capabilities.[3] From the donor's concern for the current state of the American technological enterprise, we in effect received our mandate to address contemporary issues.

But, there are contemporary issues and there are contemporary issues. I have just taken part in a week-long Lemelson symposium/exhibition, entitled *Electrified, Amplified, Deified: the Electric Guitar, Its Makers, and Its Players*. Everyone I have mentioned it to is extremely curious. 'What does it have to do with invention?', they ask. It turns out 'a great deal', as we discovered from a week that featured not only rock-and-roll concerts, interviews with guitar makers and performers and a major exhibition of vintage guitars (figure 1), but also a scholarly symposium that brought together historians, anthropologists, guitar makers and the general public.

There was a tremendous public response to these events (touts actually hawked tickets at one of them) and enormous press coverage, even for the Smithsonian. What accounted for this response? There was certainly an abundance of technological ingenuity to admire. We could have organised a fascinating science-centre-type presentation on acoustics, electronic circuits, sound reproduction and the like. But, it was context, I would argue, that made this topic particularly appealing. And the Museum was the right sort of forum to capture these broader dimensions. We brought

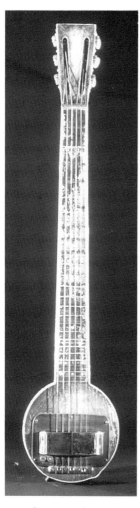

Figure 1. The Rickenbacker guitar, c.1931, is one of the earliest electric guitars. Crafted out of a single piece of wood, this particular guitar was the prototype for a commercially produced cast aluminium model, which earned it the nickname 'frying pan'.

vintage look (the same ploy used to great success by the manufacturers of Harley-Davidson motorcycles). Most of all, our audiences could learn about the critical roles of class, race, gender and social location in the upstart emergence of the electric guitar, a phenomenon of the American periphery.

Many of these humanistic aspects which fascinated the public would have been lost, I suggest, in a pure science-centre presentation of electronics or acoustics. While I have no doubt that the driving rhythms of rock-and-roll music were the main public draw, it was clear from the quality of questions and discussion that the public responded to more subtle messages as well.

Achieving this blend of technology and culture was admittedly a stretch for the Lemelson Center. In the beginning, we had started with the traditional emphasis on mechanical inventions, from automobiles to zip fasteners. But, in keeping with the broad mandate of the National Museum of American History, we have been seeking new ways to view invention and technology in a socio-cultural context.

Explaining this broadened interpretation of invention to our donor was critical to our relationship with him. A successful museum–sponsor association is always one of mutual understanding and respect. Jerome Lemelson came to us with traditional mechanical concepts of invention and a science-centre model in mind. But, though perhaps unacknowledged, his vision of technology was fundamentally contextual. His concern, after all, was not with invention per se, but with economic and cultural benefits that he saw as critical to American well-being and self-identity. In his quest to revive America's technological leadership, Lemelson in reality had a cultural message to deliver. In a sense, invention in itself was almost a side issue.

Recognising this, we succeeded in convincing him of the importance of context and history on our terms. During our negotiations, for example, we introduced him to our historical manuscript collections and, when he

together people and technology: the players, the makers and their instruments. Seeing them at work seemed to mesmerise our visitors. Among other things, the public saw the relationship between artisanship and invention, and the aesthetic impulse often driving invention. Also on display was the role of sheer play and technological enthusiasm. The reciprocity between culture and technology could be observed, for instance, in the musicians' demands for a technical fix (to compete with the loud voices of the horns in 1930s bands, guitarists sought amplification), and in their fascination with the new sonic possibilities in electronic circuits. We also saw the pull of the past with guitar makers who purposely 'antiqued' their modern products for a retro, saw the papers of Earl Tupper, the father of Tupperware, he began to understand what we were about. Instantly recognising in Tupper someone like himself, he became a convert to documentary preservation and to history.

Having revealed the essence of our successful argument to Lemelson, I must now admit that we also had to convince ourselves at the National Museum of American History of the integral relationship between technology and culture. For, in fact, one of the prime reasons for our management's interest in the Lemelson programme—and the main reason we shaped the programme as we did—was the Museum's long-standing difficulty in coming to terms with the cultural dimensions of technology and related issues of contemporary relevance.

At precisely the moment Lemelson came on the scene, the Museum was undergoing a major reorganisation designed, in large part, to grapple with these issues. The immediate spur to reorganisation was the necessity of downsizing faced by virtually every public museum today. We were moving towards a leaner, meaner, market-driven organisation. It was a matter of faith, supported by some data from audience evaluation, that the public wanted more engagement with contemporary issues. Yet, the National Museum of American History, like many museums of its type and vintage, was mired in the nineteenth century, when great international exhibitions provided the rationale and collections for the modern industrial museum. Compounding the difficulty of addressing the 'here and now' was an attitude among a vocal minority of curators that any involvement with current subjects raised political and ideological issues that threatened to compromise scholarly objectivity. The nineteenth century, at least, offered the shelter of historical distance.

Moreover, internal divisions within the National Museum of American History made it difficult to relate science and technology to societal and cultural factors. Fragmented into diverse specialties, the Museum has always found it hard to pursue interdisciplinary projects, despite the wealth of material culture crossing every conceivable boundary. Most worrisome was the schism between social history and technology curators.[4] Traditional technology exhibits tended to focus on the 'nuts and bolts', with little or no regard for the social dimensions of politics, labour, the family and other facets of everyday life.[5]

To a management uncomfortable with these lapses and internal divisions, the Lemelson programme seemed to offer a tonic and, when the Lemelson opportunity appeared, it seemed best to give it an enabling—and healing—role in the institution. We hoped to build upon the Museum's strengths, even as we strove to bring together its diverse parts to serve the new conception of the National Museum of American History.

The result was a set of programmes serving as an arm of the Museum. Provided with a modest headquarters adjacent to science and technology exhibitions and with a small core team (historians, educators, technical support and programme staff), the Lemelson Center is formally a programme under the Museum's History Department, the home of its curatorial divisions. We support the National Museum of American History's various historical programmes, and, to a lesser extent, those of other Smithsonian museums. We also fund the work of some outside historians in the realm of invention and innovation, but such support is strictly in partnership with Smithsonian staff.

Our promotional literature gives a sense of our Center's overall mission:

The Center explores the role of invention and innovation in the United States, particularly in its historical context. Drawing on the Smithsonian's vast collection of artefacts and archival materials, Lemelson Center programmes: **1** record the past, by preserving records and artifacts; **2** broaden our understanding of invention, through research, discussion and publication of ideas; and **3** engage young people in the study and exploration of invention.[6]

In my opinion, the strength and originality of our programmes reside in three sets of tensions. First, while we focus on the contemporary, defined roughly as the post-Second-World-War era, we also trace historical roots—as far back as Leonardo da Vinci if appropriate. However, pre-war projects are expected to demonstrate some relevance to current topics.

Second, while increasing public understanding is our overriding goal, we build upon scholarly foundations. More precisely, we strive to interweave scholarly and outreach activities whenever possible. We realise this ecumenical approach is unusual, as most other centres for public understanding of technology (such as the Dibner Institute in Cambridge, MA, at one end of the spectrum to Akron, Ohio's Inventure Place, on the other) cultivate niche audiences.[7] But, we felt a dual approach best suited the needs of the Museum and would serve to cross-fertilise educational and scholarly strategies. That has indeed happened. Packaging scholarship for educational programmes ensures it will reach scholars too. Moreover, the opportunity to work with children is viewed as a bonus by team members, scholars and educators alike.

Third, while all our projects revolve around a technological core, we are resolutely interdisciplinary in our quest to understand and convey the cultural contexts of invention. Our guitar programme is perhaps the most dramatic illustration of our attempts at contextualisation.

Our specific programmes include documentary projects, chiefly in the form of video and oral histories, in which inventors and innovators tell their own stories. Documented subjects range from computers and microelectronics, to synchrotron radiation, to a crude but ingenious 'cash–carry' system for handling money in a small department store, to digitised versions of carpenters' tools.

One of the most frequent complaints one hears about such electronic records is difficulty of access. The Center aims to increase public access to these and other histories through what we have dubbed the Modern Inventors Documentation project (MInD). Situated in the Museum's Archives Center, the MInD project develops databases on invention and provides an information clearing-house for inventors' papers.

The Smithsonian has been given a mandate to widen its audiences and to take its research, collections and programmes to the people. Much hope is invested in the new electronic media, from cable television to CD-ROMs to the Internet. At the Institution, the Lemelson Center has taken a lead role in experiments with electronic outreach. Supported oral history and video history projects, for instance, provide grist for a range of multimedia projects, geared to home use, school curricula and other educational programmes. The Lemelson Center's Home Page is an expanding presence on the Internet and we are supporting other Web sites as interactive stations within our exhibitions.[8] The development of a major CD-ROM on invention is currently under way.

The centrepiece of the Lemelson programme is our annual autumn symposium, *New Perspectives on Invention and Innovation*, which brings together historians, practitioners, entrepreneurs and the public, for a week of lectures, discussions, workshops and exhibitions. For these high-profile events, we have focused on interdisciplinary topics of known public appeal. How we selected the topic of the electric guitar is instructive of our process. Keenly aware of the tremendous public interest in the modern electric guitar, among both adults and teenagers, curators in the Museum's music section proposed a programme on the topic to the Center. Seeing vast potential in their proposal, we suggested expanding it into our annual symposium. The result was, as intended, an exploration not only of invention but of a key facet of US cultural history. Last year's symposium, entitled *The Inventor and the Innovative Society*, explored the social and cultural contexts of Leonardo da Vinci, Thomas Edison and innovators in 'Silicon Valley'. In 1997, the symposium will examine technologies of colour from the rise of the dye industry, to photography and film, to the latest in digital colour. We plan to feature prominent fashion designers and cinematographers to dramatise in lay terms the broad cultural implications of this technology. As these programmes illustrate, our contemporary emphasis in no way precludes attention to earlier eras. Rather, we aim for a dialogue between past and present that will provide historical perspective on current issues. Other Lemelson conferences of a lesser scale have explored such current topics as technology policy, the rise of molecular medicine and architectural innovation—all of which featured first-hand accounts by current practitioners.

Whenever feasible, we link our scholarly proceedings with wide-reaching public programmes for both adults and children. Thus the electric guitar symposium included concerts and musical demonstrations by inventors and players. A public interview in our *Portraits of Invention* series on Les Paul, a premier inventor and performer on the electric guitar who gave the instrument its modern form, allowed us to bring invention to life. It also enabled us to draw in people who may have had no prior interest in technology and innovation. Developing these broader audiences is key to raising the level of public literacy about invention and technology. The publicity that comes so easily with such programming dramatically increases the visibility of the inventive process.

Our most successful evocation of the 'here and now' has been our *Innovative Lives* series: monthly lecture/demonstrations that introduce students to inventors and entrepreneurs who share personal stories of invention. Recent speakers included Chuck Hoberman, a follower of Buckminster Fuller and inventor of the Hoberman Sphere, whose presentation of foldable structures complemented our symposium on architectural innovation; Stephanie Kwolek, inventor of Kevlar fibre, who gave a woman's perspective on scientific careers; and James McLurkin and Thomas Massie, two young MIT graduate-student entrepreneurs, who demonstrated their recent computer innovations, including experiments with miniature automata. The *Innovative Lives*

series has been particularly successful in recruiting minority and women student audiences and speakers, who show that careers in inventing today are open to both sexes and all races (figure 2).[9] Selected *Innovative Lives* programmes are taken around the country and the world through 'electronic field trips,' one of which recently connected American students at the Smithsonian with British students at the Science Museum.

The *Innovative Lives* series has been perhaps our most successful outreach programme to date, enthusiastically received by students and teachers, who use our supplementary curriculum materials to prepare class lessons around the event. From our surveys of participants, we know we are fulfilling a real need. These events naturally require a great deal of advance preparation. The Center's education staff work closely with teachers, cultivating partnerships with schools and often providing buses to bring in students. Follow-up with teachers after the event is just as intensive.

A merger of interdisciplinary scholarship and outreach can also be seen in our Intern and Senior Fellows programmes. Internships bring college and graduate students to the Smithsonian to learn professional museum skills. Senior Fellows—of whom we support two to four per year—bring their own perspectives to the programme, pursuing historical research or educational projects. Recent Senior Fellowship projects include work on African–Americans and invention by Bruce Sinclair, which will result in a *Needs and Opportunity* volume, plus a companion book of sources for high school teachers.

Summing up

Experience shows that all our audiences—interns, scholars, practitioners, the general public—are excited by seeing invention and technology in context. I don't necessarily mean 'context' only in a sophisticated sense. For the uninitiated public, it may be as simple as saying that people enjoy seeing other people—

Figure 2. A radio is dismantled in a search of raw materials for new inventions as part of Innovative Lives, *a successful outreach programme*

and preferably *live* people. For the scholar, it may simply amount to juxtaposing invention with social phenomena, although the Center also entertains and encourages more complex sorts of analyses.

Placing invention and technology in cultural context has proved to be an effective technique for introducing broad audiences to a difficult subject. It is also the responsible approach to contemporary topics in technology. It is, after all, with the social and cultural effects of the computer, the Internet, new medical technologies, genetically engineered organisms and the like, that our public must grapple on a daily basis. Public understanding of these issues is rapidly becoming a requisite of citizenship.

Admittedly, interdisciplinary perspectives can lead to confusion and ambiguity. When I announced this approach to the staff of the Museum, the social and cultural curators were delighted—and all too creative in posing questions such as: can we regard the Republican party's campaign ploy, the 'Contract with America,' as an invention?; how about the Bill of Rights?; and so on (they have continued to be extremely inventive in applying for the Center's funds). Not wanting to lose the basic identity of invention, we finally arrived at the rule that cross-disciplinary projects would need to begin with a technological core. How far they could then move away from that core would be evaluated on a case-by-case basis. With our symposium on the electric guitar, I think we were pushing, but still within, the boundaries, stressing as we did the role of technical innovators, some of whom were not musicians at all.

In order to retain a recognisable core, where do you draw the line? With the electric guitar, we thought a great deal about boundary questions. How much emphasis should be placed on the music? On the lives of makers and players? On the electronics? On the instruments? On manufacturing techniques? For the guitar, the combination was critical. As I noted before, people were drawn first to the music, but they revealed a deep fascination with the technical ingenuity on display. This may be a phenomenon peculiar to American audiences, who harbour cherished myths of Yankee ingenuity, but I suspect not, given the universal appeal of the instrument. For the scholars involved in the symposium, the relations between culture and technology were equally compelling and complex.

For some internalist historians, as well as for some hard-nosed inventors, this contextualising only detracts from an appreciation of the technology. I have argued the opposite. It is my belief that broadening the contexts for technology and invention to include the arts, music and other fields makes them far more important than some people, including some specialised scholars, ever suspected.

Notes and references

1   This may be more of an issue for Americans than Europeans. See Post, R and Molella, A, 'The call of stories at the Smithsonian Institution: history of technology and science in crisis', *ICON*, 3 (1997), in press.

2   Jerome Lemelson is a fascinating figure in his own right. For portraits, see Kernan, M, 'Around the mall and beyond', *Smithsonian Magazine*, 27 (1996), pp20–22; Maloney, L, 'Engineer of the Year, Jerome Lemelson', *Design News* (6 March 1995), pp70–72; Baily, A L, 'The inventor's latest: a new Foundation', *The Chronicle of Philanthropy*, 7 (10 August 1995), pp8–10.

3   Lemelson has initiated programmes at the Massachusetts Institute of Technology (MIT), which awards an annual Lemelson–MIT prize for invention; at the University of Nevada; and at Hampshire College in Hampshire, MA, which also administers his national programme, National Collegiate Inventors and Innovators Alliance.

4   For the split between the science and technology areas, on one hand, and the social and cultural on the other, see Molella, A, 'Tilting at windmills', *Technology and Culture*, 36 (1995), pp1000–06.

5   The ideological ramifications of this approach are explored in Molella, A, 'The Museum that Might have Been: The Smithsonian's National Museum of Science and Industry', *Technology and Culture*, 32 (1991), pp237–63. Not that visitors necessarily feel they are missing anything: faced with the choice of admiring a huge steam locomotive or reading labels about railway labour, what do you think the average visitor will do?

6   From our general information brochure, available from the Lemelson Center Office, National Museum of American History, Smithsonian Institution, Washington DC, 20560, US.

7   Schwarz, F D, 'A hall of fame for inventors', *American Heritage of Invention and Technology*, 11 (1996), pp6–7

8   Lemelson Home Page: http://www.si.edu/lemelson

9   Our presenters included Ellen Ochoa, the first female Hispanic astronaut. African–American speakers have included, besides James McLurkin, Hal Walker, engineer and entrepreneur, and astrophysicist George Carruthers.

Collaborate or stagnate?

Bridging the gap

Heather Mayfield and Peggie Rimmer

One of the world's leading research laboratories has forged a close relationship with a museum 500 miles away.

At first glance, the European Laboratory for Particle Physics (CERN) in Geneva does not have much in common with a science museum. CERN is one of the world's leading centres for the investigation of the ultimate constituents of matter and for finding out how the universe works—its business is very much 'cutting edge' science, with no room at all for out-of-date ideas or equipment.[1] Science museums, on the other hand, are most famous for their presentations of the past, usually with very little attention paid to the contemporary.

Given these differences, it is perhaps surprising that CERN has fostered a close relationship with a science museum, especially as that museum is nearly 500 miles away. The museum concerned is the Science Museum, London.[2]

In this paper we shall describe, mainly from our own points of view, how the collaboration began and how it has been nurtured to the mutual advantage of our institutions. We believe that this personal story illustrates how a 'big science' laboratory can work fruitfully with an institution whose primary purpose is to promote the public understanding of science—the rewards of the collaboration are many and various, as we shall describe.

Why collaborate?

Peggie Rimmer (PR), CERN
CERN is a facility run by just under 3000 staff members for the benefit of about 6500 users, representing over 80 nations from more than 500 universities and institutes. CERN's annual budget of some 900 million Swiss francs is provided by 19 European countries.

As a publicly funded organisation, CERN has long recognised its duty to explain its mission and achievements. The Laboratory was a pioneer in 'hard science' exhibitions, dealing with an area in which the lay person has little understanding of the subject matter or the way in which it is investigated.

CERN has the world's only travelling exhibition on particle physics, which made its debut in Hanover in 1978 and has been 'on the road' ever since. The outreach potential of this exhibition is particularly important in spreading awareness of CERN and particle physics in distant member states such as Finland, Hungary and Greece. In 1989 CERN created *Microcosm*, an on-site exhibition about particle physics which looked at its history, but concentrated on modern ideas and on the experimental work being done with CERN's particle accelerators. Each year, this exhibition attracts some 25,000 visitors.

Both exhibitions are a credit to their designers, but they have recently been somewhat overtaken by events. The 'lecture notes on the wall' approach (figure 1) looks dated compared with the exciting appearance of many of today's science centres. Ironically, even their interactive computer programmes, at one time state-of-the-art, must now compete with material at least as good that which can be readily located on the World Wide Web, itself a CERN invention.[3]

As funding for fundamental research becomes more contentious,[4] governments and taxpayers are demanding a clear and convincing view of the benefits. In trying to communicate this view we are aware that high-quality, visually stunning, multimedia presentations are a part of everyday life. So CERN turned to experts from the modern museum world to help improve the image and impact of its exhibitions.

Heather Mayfield (HM), Science Museum
Science museums are known for their ability to present historical material. The best of these museums have collections of unique and sometimes beautiful objects, together with staff who are experts on their provenance and place in history. A typical gallery, focusing on the historical development of a theme, may take years to gestate and will, in all probability, not be replaced for a decade.

a few specialist contemporary science displays—they have no choice but to look outside for advice and assistance.

At the Science Museum, we have gained a good deal of experience with the presentation of contemporary science and technology through our *Science Box* exhibitions.[5] In preparing them, we have had to rely on the knowledge and enthusiasm of scientists to produce accurate, stimulating and innovative exhibitions. That said, we have encountered some challenges in working with these experts.

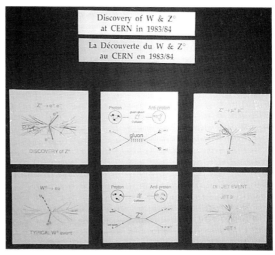

Figure 1. CERN has wonderful objects to display like the Big European Bubble Chamber (left), but the explanations are sometimes difficult to decipher (above)

Increasingly, however, science museums are responding to the public interest in contemporary science and technology. This places quite different demands on exhibition developers: for example, they have to ensure that their material is up to date by obtaining objects and expertise directly from experts in academia and industry. No science museum could ever hope to have the resources to provide in-house expertise for any more than

Many scientists are keen to tell people what they do and why they do it—future funding for their work might depend on how well they do this. However, the ability to be an outstanding scientist by no means guarantees an ability to be an expert communicator, nor an ability to come up with ideas for exhibitions.

Some scientists take some convincing that simplification is not trivialisation and that the average level of our visitors' scientific understanding is about that of a typical 15-year-old, not a high-flying degree student. Moreover, when people come to a museum or science centre, they expect to be entertained as well as educated. Scientists have to accept that the urge to educate a lay audience can no longer be divorced from the need to attract visitors, for example a family on a Sunday afternoon trip, who are window-shopping for the most enjoyable experiences on offer.

Overall, our experiences of working with leading scientists have been immensely rewarding for us and, we gather, enjoyable for them too. It is always important to bear in mind that ultimately it is our responsibility to communicate effectively with the public: as Winston Churchill remarked, 'Scientists should be on tap but not on top.'[6]

How did we collaborate?

PR, CERN
In 1995, when I took over responsibility for CERN's exhibitions, I knew that they needed a facelift and that the necessary know-how was not available in the community of particle physicists. I therefore decided to invite an exhibition expert to have a short, sharp stab at putting us on the road to improvement. The question was: who? I started networking in the science exhibition world by visiting several major centres and attending meetings such as those organised by the European Collaborative for Science Industry and Technology Exhibitions (ECSITE). I was looking for someone with experience in both permanent and travelling science shows, who could work within the culture at CERN. The ideal fit proved to be Heather Mayfield.

HM, Science Museum
When Peggie contacted me, I had been manager of the *Science Box* series of exhibitions for two years and been responsible for producing eight exhibitions in a row, without a break. I felt the need for a change of scenery and to develop my skills in a different environment.

During my career as the *Science Box* manager, I had become curious about how leading-edge research is carried out and was keen to gain experience in an environment in which pure scientists have the upper hand. The opportunity to work at CERN for a few months therefore suited me perfectly, especially as I could foresee the opportunity of setting up a collaboration between the Science Museum and CERN in the production of a mooted exhibition on the Big Bang.

The Museum granted me three months' leave of absence, and in May 1995 I went on a placement in CERN's Communication and Public Education (COPE) Group, headed by Peggie. My principal task was to evaluate CERN's travelling and permanent exhibitions and to produce a report on the way forward for their exhibitions unit.

What did we get out of the collaboration?

PR, CERN
In September 1995, Heather presented her report to CERN's top management, outlining the state of the exhibitions and offering some down-to-earth recommendations for improvement. It was a sobering document. Perhaps the most important chapter contained quantitative data collected in the first-ever visitor studies carried out in the *Microcosm* exhibition centre. These data enabled us to gauge, for example, how effective the displays were at getting over their messages and how much the visitors were profiting from their time in the exhibition. This approach appealed to an audience of physicists and engineers—it engendered confidence in the results, which demonstrated unequivocally that changes were needed.

Support for the renewal of CERN's exhibitions was soon forthcoming and I was able to employ a young science communicator. Two projects were approved in early 1996, each for a three-year period. The first is to produce a new, modern travelling exhibition with both the design and building contracted out to professionals. The second project is to refurbish the basic amenities at *Microcosm* to make the centre a more lively and attractive element in CERN's visitors' programme. I hope that this will lead to a complete refit of the *Microcosm* exhibition, with a millennium exhibition worthy of CERN's twentieth-century achievements and twenty-first-century programmes.

HM, Science Museum

During my 12 weeks at CERN, I had access to staff at all levels within the organisation. One of my tasks was to work with the exhibition staff and scientists in the laboratory to find out who they understood the visitors to their exhibitions to be. By focusing on the audience rather than the content of any exhibition, I was able to look at the whole exhibition service and consider what they might do in the future rather than dwell on what had been produced in the past.

Some studies had already been carried out by the CERN exhibition staff. They had worked out their cost per visitor to the *Microcosm* exhibition and were keen to boost visitor figures and to revamp the centre to reach more of the public. An extensive training programme was already under way for the guides who take visitors around the CERN site and one of the major parts of this training involved learning how to explain the purpose of the rather recondite experiments to non-specialists.

Working at CERN gave me the chance to see a range of activities that take place in an international scientific institution. For the first time I was able to determine what scientists in a multinational laboratory want to derive from presenting their work through exhibitions. I had opportunities to observe experiments, take part in and contribute to the guides' training programme, attend some of the lectures provided for the summer students, and get to grips with the problems of developing exhibitions and resources in at least two and preferably four languages simultaneously (some 15 languages are spoken in CERN's European

user community alone). With the proposed 'Big Bang' *Science Box* in mind, I was particularly interested in any work within Peggie's group that examined the interpretation of particle physics and cosmology for a lay audience.

By April 1996, the Science Museum Temporary Exhibitions Unit and CERN physicists had begun to collaborate on the development of a small exhibition on the Big Bang, and had secured additional funding from the UK's Particle Physics and Astronomy Research Council. After the team had worked on the material for four months, including two visits to CERN, the exhibition was opened at the Science Museum in September with an inaugural lecture by CERN's Director General, Christopher Llewellyn Smith, attended by about 50 leading members of the UK particle physics community. His presence was instrumental in our securing excellent media coverage for this opening—coverage that also enabled him to put the case for funding CERN to a wide audience in the UK through interviews he gave for radio, television and newspapers. This *Science Box* was at the Museum until January 1997 and moved to CERN for six weeks in February 1997, before its UK tour, funded by the UK government's Department of Trade and Industry.

Results

HM and PR

Staff from the Science Museum and CERN have been working together since the earliest days of the Laboratory in the 1950s. Past collaborations have been between individuals, with the Science Museum generally being in search of expertise, objects, information or illustrations. Our collaboration has consolidated working relationships at several levels and has opened up many new channels of communication between our institutions. Curatorial colleagues at the Museum are benefiting from the new contacts and CERN's perception of its public relations is undergoing a sea-change.

In the summer of 1995, John Durant, an Assistant Director of the Science Museum and Professor of the Public Understanding of Science at Imperial College of Science, Technology and Medicine, London, was invited to CERN to give a seminar entitled 'Why should the public understand science?'. This raised consciousness of the issues in CERN to the extent that, from 1997, the public understanding of science will be a regular item in CERN's academic training programme. The first lecturer will be Nick Russell, Senior Lecturer in Science Communication at Imperial College.

In 1996, CERN welcomed its first-ever summer student in science communication. Using the Imperial College network once again, a young physicist/communicator was invited to make an appraisal of objects and hands-on methods suitable for particle physics exhibitions. We hope that such a placement becomes an integral part of CERN's summer student programme. Also, towards the end of 1996, Roger Miles, formerly of the Natural History Museum, gave a short course on exhibition design to CERN's exhibition team.

Richard Gregory, the distinguished neurophysiolgist and science-centre luminary, once observed that 'it is remarkable how little science there is in traditional science museums'.[7] He might also have added that it is remarkable how few outreach activities are undertaken by traditional science laboratories. The essence of our collaboration has been for each of our institutions to offer its strengths for the other's benefit.

Notes and references

1 Details of CERN's work may be found in its annual report, a copy of which is available from the Communication and Public Education Group, CERN, 1211 Geneva 23, Switzerland. The World Wide Web address is http://www.cern.ch/

2 Details of the Science Museum's work may be found on the World Wide Web at http://www.nmsi.ac.uk/

3 Berners-Lee, T, 'The World-Wide Web', *Commun. ACM*, 37 (1994), pp76–82; for a general history of CERN see Hermann, A, Krige, J, Mersits, U, Pestre, D and Belloni, L, *History of CERN* (Amsterdam: North-Holland, vol. 1, 1987 and vol. 2, 1990).

4 For a recent discussion of the issues see the editorial article 'An accelerator worth fighting for' in *Nature*, 382, p376.

5 Ward, L, this volume, pp83–90

6 Churchill, R, *Twenty-one Years*, epilogue (London: Weidenfield and Nicolson, 1965); this quotation has also been attributed to Walter Elliot (1888–1958).

7 Gregory, R L, 'Turning minds on to science by hands-on exploration: the nature and potential of the hands-on medium', *Sharing Science: Issues in the Development of Interactive Science Centres* (London: The Nuffield Foundation on behalf of COPUS, 1989), pp1–9

Outreach from an international laboratory: a task out of reach?

Dominique Cornuéjols

One international research facility has not found it easy to persuade museums and science centres to collaborate.

At the foot of the French Alps in Grenoble lies one of Europe's leading research centres, the European Synchrotron Radiation Facility (ESRF, figure 1). Here, several thousand scientists from all over the world use ultra-brilliant X-rays to carry out research into a wide range of fields including chemistry, physics, biology and materials science. All over Europe there are hundreds of museums and science centres, many of them interested in communicating contemporary science. Yet no strong links exist between them and the Facility—indeed most museums and science centres seem unprepared to collaborate.

In this paper I want to look at the scientific work of the Facility and its outreach activities before examining the potential for collaboration with museums and science centres. I suggest that more collaboration could help to meet increasing public interest in understanding contemporary challenges in science and technology.[1]

What is the European Synchrotron Radiation Facility?

The basic purpose of the Facility is to use synchrotron radiation (high-energy light) to

Figure 1. The European Synchrotron Radiation Facility in Grenoble

study matter. But what is this and how do we generate it? Any electrically charged particle gives off radiation when it is accelerated. For example, if high-energy electrons are made to move in a curved path they emit radiation, generally known as synchrotron radiation (X-rays are part of this radiation). This is how the Facility generates the radiation that forms the basis of all its experiments: it accelerates electrons to very high speeds, close to the speed of light, and generates synchrotron radiation,[2] which is given off, rather like water droplets flying off tangentially from a spinning bicycle wheel. It is this radiation that is the key tool in experiments at the Facility.

The Facility is multinational, and this has a political and economic significance: it shows that scientists from different countries can work together and are stronger than they would be if they worked in separate institutions. For this reason, today's research increasingly involves multinational collaboration.

The user-oriented approach of the Facility is also a new challenge. A huge piece of scientific equipment is at the disposal of scientists from all over the world. This encourages researchers to be mobile and it is an incentive for openness towards new techniques and developments.

Artificial categories as usually defined (e.g. physics, chemistry, biology) do not necessarily correspond to the reality of condensed-matter science, which is a vast interdisciplinary field. The Facility is an ideal environment for scientists from different sub-fields to meet and discuss their common interests.

A further characteristic of the Facility is that it is highly instrumental. Scientific results depend directly on the performance of the equipment and this means that scientists must have access to state-of-the-art instruments if they are to work at the forefront of research.

Some of the more interesting fields of science studied at the Facility include molecular biology, imaging and materials science, which are shown in table 1.[3] In addition to these, the emerging field of high-pressure research is opening up new possibilities, for instance in

synthesising diamond, developing ultra-hard ceramics for the aerospace industry, studying semiconductors under pressure and finding new means of sterilising food.

Spreading the message

The main responsibility of the ESRF Information Office is to report the research activities of the Facility to the scientific community. A secondary task is to spread the word to the non-specialist public. The Office presents the public face of the Facility, by being the first point for contact with journalists, politicians and representatives from other institutions. It is our task to produce material such as brochures, posters, videos and CD-ROMs that make the scientific work of the Facility accessible to lay people. We welcome between 3000 and 4000 visitors each year from all over the world, including students, scientists, engineers, delegates to international conferences in Grenoble and the general public. Unfortunately, we have to refuse half the requests for visits, especially from secondary schools, because we do not have enough personnel.

All these tasks are performed by only two people: a secretary and myself. Additional activities such as 'Science en Fête', the French National Week for Science, exhibitions, open days, and information campaigns in schools and universities, can be developed only when we are able to employ an extra person on a fixed-term contract, with financial help, usually from the European Union.

Outreach from partnership

The most challenging outreach activities undertaken by the Facility have been carried out within the framework of the European Weeks for Scientific and Technological Culture set up by the European Commission Directorate General XII (DGXII). The central purpose of these weeks, which have so far taken place in the third week of November 1993–96, has been to bring science and technology closer to the

Table 1. The applications of synchrotron radiation to molecular biology, imaging, and materials science

Molecular biology	*Imaging*	*Materials science*
Molecular biology is the field in which the most dramatic progress has been made in recent years using synchrotron radiation. In this field, scientists look at the molecular structure of a protein or a virus to enhance understanding of biological functions and thus develop new drugs, design new biomaterials and improve industrial iocatalysts.	Until recently, the use of X-rays in imaging (such as in radiology) has been limited to looking at 'hard' tissues (bones, tumours). 'Soft' tissues (muscles, organs) absorb radiation uniformly, making it impossible to see the outlines of these tissues using the usual absorption techniques.	Hundreds of materials are studied at the Facility.
		Catalysts are being studied in situ by the petrochemical industry, in order to understand better what happens during catalysis and so design more efficient catalysts.
X-ray crystallography is the most important and most frequently used method for the determination of the structure of protein molecules. The X-ray beam at the Facility, which is narrow and of high intensity, is ideal for the crystallography of biological material.	At the Facility, the beam is coherent, i.e. very similar to a laser beam, but in the X-ray range. This allows for 'phase contrast imaging' which produces striking images showing details that do not appear at all in the simple 'absorption mode'.	*Polymers*, artificial and natural, are being studied under different conditions, especially at different stages of tensile stress. Spider web fibres, for example, show exceptional strength compared with steel.
		Magnetic layers, semi-conducting devices and growth processes on surfaces can be observed with an unprecedented precision.

European public through the 20-odd events that have been presented each year.[4]

The organisers at DGXII have gone out of their way to involve the 'big' laboratories in Europe and to promote collaboration between different European institutions. The Facility has benefited considerably from this policy which has provided us with excellent opportunities to address some of the interest from the public. It has also enabled us to foster some new and fruitful collaborations. Detailed below are the activities that the Facility has organised as part of the European Week.

At the Heart of Life (1993)

For the first European Week, we organised an open day and *At the Heart of Life*, a conference on biology, in collaboration with neighbouring institutes, the Institut Laue-Langevin and the European Molecular Biology Laboratory (EMBL). Admission was by invitation only: we invited around 500 secondary school pupils to visit the three institutes. We also advertised the open day in a weekly newspaper, *Les Affiches de Grenoble*: on the front page and three inside pages we invited the readers to participate on a first-come, first-served basis. *Les Affiches* was distributed on Friday afternoon as usual. But, by the following Monday, more than a thousand responses had been received, and some readers came directly to the newspaper's office on Friday night or Saturday rather than sending their reply by post. *Les Affiches* had never before experienced this level of response, and we were delighted to organise two open days in order to satisfy the demand.

Materials for the Future (1994)

In 1994, a similar event was organised called *Materials for the Future*, which consisted of two open days and a conference. It, too, was organised by the ESRF Information Office, and a similar level of success was achieved. The most innovative feature was the co-production of a pilot CD-ROM with Daresbury Laboratory, a synchrotron radiation facility in the UK. The CD-ROM was designed to reach a broad audience and was especially attractive to 15- to 20-year-olds during the open days, during which the CD-ROM could be accessed. Other partners for the pilot CD-ROM development were academic research groups and several industrial companies involved in synchrotron radiation research in Europe. Unfortunately, this CD-ROM project could not be completed, owing to lack of resources, although we intend to relaunch it in 1997.

100 Years of X-rays . . . and the Future? (1995)

The 1995 centenary of the discovery of X-rays by the German physicist Wilhelm Röntgen[5] was an ideal opportunity for us to celebrate the work of the Facility and X-rays in general. We organised a competition for young Europeans (aged 15–25), who were asked to 'imagine an application for X-rays in the future'. Forty thousand leaflets were distributed in secondary schools and universities in 10 different European countries, in English, German and French. The competition was also announced in newspapers and scientific magazines. Industrial groups were partners, as was the Science Museum, London. A 52-minute television documentary on the history and the present use of X-rays was co-produced by the Facility and broadcast on three channels.

Our attempts to collaborate with European science centres and science museums were woefully disappointing. About a year before the event, we looked into what was being organised in connection with the centenary by other institutions in Europe. The Röntgen Museum in the discoverer's birth place of Lennep, Germany, and Würzburg University, where he first observed X-rays, both had preparations for new exhibitions well under way and were not particularly interested in collaboration. Apart from some medical institutions dealing with radiology, most science centres were not the slightest bit interested in the centenary.

Faced with such indifference, we decided to concentrate on our own projects—the competition and the television documentary as described above. The winners of the competition (12 young Europeans) were selected by three juries (scientific, journalist and industrial juries) and then invited to attend three 'discovery days' at the Facility in November 1995.

We were more successful in establishing collaboration with the press (*Physics World, La Recherche, Natuur & Techniek, Die Welt, Science & Vie* and *Science & Vie Junior*), who announced the competition and were members of the journalists' jury (some of them even sponsored the event), as well as industry: Glaxo for pharmaceuticals, L'Oréal for cosmetics, BASF for petrochemicals, Pechiney for metallic alloys and HTAS (a Danish company) for catalysts. In particular, a team from L'Oréal agreed to come and carry out a live experiment in front of the competition winners on the Facility's beamline during the Week.

We also worked with the Science Museum, London, who sent one of the resident actors who had developed a role based on the experience of the only journalist who was fortunate enough to have secured an interview with Röntgen after he made his discovery.[6] The innovative performance by the actor at the Facility's centennial celebrations was much enjoyed by his audience, who were stimulated to go beyond the details of the discovery and to question the relationship between scientists and journalists.

We contacted the museums and science centres in Europe for a second time when we were ready to announce our competition. We presumed that science centres were ideal places for this distribution and that they would be keen to work with us, but we were mistaken. Some refused to distribute the leaflets, on the pretext that it was 'contrary to policy'. A few reluctantly agreed to help and we sent them parcels containing 1000 or 2000 leaflets, but we do not know whether these were distributed as we did not receive a single response.

Science Ambassadors (1996)

The *Science Ambassadors* project was organised in collaboration with the European Laboratory for Particle Physics (CERN), the European Space Agency (ESA), the European Molecular Biology Laboratory (EMBL) and FUSION (research and development programme of the European Commission, including JET, the Joint European Torus).

During two days of the Week, 20 post-doctoral researchers from these European research institutions were sent to the secondary schools where they had studied, in 16 European countries, to speak about their jobs in an exciting scientific environment, and to show the youngsters the opportunities and careers available to young scientists nowadays. After this direct contact with pupils, students and teachers, the Science Ambassadors met in Grenoble to talk about their experience in the schools and try to answer the question: 'What can be done to improve the links between the European research centres and the educational system so that pupils can fully benefit from the high profile of these centres?' Heads of communication from research centres were present, as were representatives from other institutions, such as schools and universities, the European Commission and the European Ministries of Research and of Education.

Outreach: whose responsibility?

Because a large part of the funding for research institutes comes from public money, it is only proper that taxpayers are told how their money is used. Surprisingly, however, many scientists do not seem to be completely convinced of this. For most scientists, communication simply means publication in a peer-reviewed scientific journal. The general public is far from their minds, and they are not willing to spend their time on outreach communication. They argue that they are not trained for this task and that their competence is better employed in carrying out research. So, if it is not their job, whose job is it?

The development of communication services within scientific institutes is only part of the answer. The main task of the Information Office at the Facility is to convey information to a wide community of users who themselves are scientists. Very few resources are available for outreach communication and, until recently, this kind of communication was certainly not considered as a priority by scientists. However, we are now observing a change in attitudes. At a time when budgets in many countries are being cut at a national level, where there is competition between all branches of economic interest, and when lobbying has become the norm, scientists often appear to be isolated and powerless. If the public and the decision makers are not aware of the necessity of basic and applied research in our society as a medium- and long-term investment, why should they fund it? Even for scientists, nowadays, the importance of presenting scientific discoveries in an understandable way for lay people is obvious.

If the Facility is to present its findings to the European public, there is no doubt that it needs to work closely with the media, and notably with museums and science centres. These institutions have traditionally concentrated on the past and have not seen contemporary science and technology as a priority— quite why this is has never been clear.[7] If a change is to occur, and modern science is perceived to be important, then the Facility would be pleased to respond by entering into more collaborations.

Establishing partnerships

The Facility is very interested in promoting collaborations, but we lack resources for outreach communication. Moreover, it would not be easy to defend the practice whereby public money given for research is instead used for communication.

For their part, science centres receive public money for developing awareness of the public of science and technology, but their survival is linked to their visitor numbers, and thus their choices are most often driven by public demand (or money-driven, as in the case of sponsorship by industry). So where is the common ground? The fascination for science itself should be the strongest common interest: our visitors seem to be so enthusiastic about what they see at the Facility, one would expect it to be the same for visitors to a science centre.

The ESRF Information Office would be very happy to send information about our work to European museums and science centres, to invite them to visit the Facility and to discuss a possible partnership in the future. Museums and science centres that are interested in collaborating with us will find that they are knocking on an open door.

Notes and references

1 Wilson, A, 'Outreach is . . .', in *Sharing Science: Issues in the Development of Interactive Science and Technology Centres* (Oxford: Nuffield Foundation, 1989), pp29–33

2 Feynman, R P, *Lectures in Physics*, 1 (Reading, MA: Addison Wesley, 1963), section 34.3

3 Modern applications of X-rays, many of which are studied at the ESRF, are well reviewed in Michette, A and Pfauntsch, S (eds), *X-Rays: The First Hundred Years*, (Chichester: Wiley, 1996). A more detailed review of the Facility's work is given in the ESRF publication *Highlights 1995/1996*, which is available from the Information Office, BP 220, F-38043, Grenoble, France.

4 Details of the events that have been organised to date in the European Weeks are available from the European Commission, DGXII (Science, Research and Development), Brussels, Belgium.

5 Glasser, O, *Wilhelm Conrad Röntgen* (San Francisco: Norman Publishing, 1993)

6 Dam, W H, 'The new marvel in photography', *McClure's Magazine*, 6 (1896), pp403–15

7 Finn, B S, 'The Museum of Science and Technology', in Shapiro, M S (ed), *The Museum: A Reference Guide* (Westport, CT: Greenwood Press, 1990)

Fostering collaboration: Wellcome Trust initiatives

Laurence Smaje

The Wellcome Centre for Medical Science is keen to fund collaborative initiatives that promote the public understanding of contemporary medical science. What are the key issues in setting up these partnerships?

In 1990, the Wellcome Trust, now the world's largest medical research charity (see overleaf), decided to establish a division dedicated to explaining the nature of medical research to the general public. The Wellcome Centre for Medical Science, as it came to be called, was not to be a public-relations exercise but had as its main mission to explain the process and the products of medical research. An explicit part of this mission was the recognition that modern biomedical research would lead to major changes in people's lives, and that decisions on the nature and direction of that research were too important to be left to scientists, doctors, industry or politicians alone but should be taken by society as a whole. This is easy to say but difficult to realise, and it was appreciated that it would be necessary to develop a long-term multifaceted approach.

What kinds of activities does the Wellcome Centre support?

The Wellcome Centre set out to complement rather than compete with other well-established bodies already involved in the public understanding of science movement. Accordingly, we help support national initiatives whose mission is similar to our own, such as the British Association for the Advancement of Science and the annual International Science Festival which takes place in Edinburgh. We try to help improve the overall quality of science communication in several ways, including providing bursaries for scientists studying on postgraduate courses in science communication,[1] and we provide funding for research into public understanding of science.

The Wellcome Centre also undertakes activity on its own account. We have an award-winning public exhibition, *Science for Life* (figure 1), at our headquarters on Euston Road, London and a well-developed programme for senior-school pupils and their teachers including lectures, seminars and hands-on workshops using molecular biological techniques such as DNA fingerprinting. In addition, a number of initiatives have been developed for the general public, including

Figure 1. Part of the Wellcome Centre's Science for Life *exhibition*

Henry Wellcome and the Wellcome Trust

The Wellcome Trust was set up in 1936 under the will of Sir Henry Wellcome (figure 2), at the time of his death the sole owner of an international drug company, the Wellcome Foundation Limited. Wellcome was born in 1853 of poor parents in Wisconsin, US, and his life story is a fascinating history of public success and personal failure.[2] He came to Britain in 1880, joining another American expatriate Silas Burroughs, and together they established Burroughs–Wellcome. Burroughs died in 1895 and Wellcome was able to take over the company as sole owner. Following Wellcome's death in 1936, the share capital of his company was vested in five Trustees: three scientists, including the Nobel laureate Sir Henry Dale, and two businessmen. Wellcome specified that the profits of the company were to be placed in a fund which was to be used to support 'research . . . bearing on medicine and allied subjects which have or at any time may develop an importance for scientific research which may conduce to the improvement of the physical conditions of

Figure 2. Sir Henry Wellcome, founder of the Wellcome Trust

mankind'.[3] It remained thus until 1986, since when the Trust has gradually sold its shares in the company, the last major tranche being sold to Glaxo in 1995 thereby forming Glaxo–Wellcome, one of the largest drugs companies in the world. The net result of this is that the Wellcome Trust now has assets in excess of £8 billion invested widely internationally and is thus as financially secure as is possible. This financial security brings with it an unprecedented degree of independence.

There are now three main sections of the Trust: the Science Funding arm which spends the major part of the Trust's income, the Wellcome Institute for the History of Medicine (the history of medicine being a subject of special interest to Henry Wellcome) and the Wellcome Centre for Medical Science. This was established in 1990 with a view to explaining the need for and the nature of medical research to the general public so as to help create a climate in which medical science could flourish.

lectures and debates, and we have created a small exhibition called *Genes Are Us*, in collaboration with the Medical Research Council.[4] *Genes Are Us* travels out to the public instead of expecting them to come to us.[5] These activities are created or managed by a core staff of eight who are wholly involved in public understanding of science activities. In addition, there are another 52 staff in the Wellcome Centre involved in other projects, some of whom are primarily supporting the medical research community, and others whose work is a mixture of both activities. Overall, about 50 per cent of the work of the Centre is concerned with public understanding of science and our total budget is £4.5 million per annum.

Funding of public understanding of science initiatives by the Wellcome Trust

The Centre's motivation for providing funds to others to undertake work commissioned by us or in response to their own proposals, include:

- the realisation that we do not have a monopoly on good ideas—there is more creativity outside any one institution than inside
- access to special skills, expertise and/or equipment
- access to target groups not otherwise readily obtainable
- economies of scale

In 1995, the Wellcome Trust spent close to £300 million on biomedical research. In consequence, it is widely believed that the Trust is so rich that it is casting about trying to think of ways of spending the money. This is by no means the case. The Trust can fund only about 30–40 per cent of eligible requests, a proportion comparable to that of the Medical Research Council.

In the area of public understanding of science, the Centre is also seen by some as a milch cow, perhaps because applicants have in mind the total budget of the Trust rather than that for external public understanding initiatives. In fact, we have a grants budget of only £650,000 per annum—significant, but by no means a bottomless pit. As with biomedical research, applications for these funds are sent to external independent referees for their opinions and these are borne in mind, along with other considerations, by an Advisory Committee which makes the final recommendations.

Focusing activity

The Centre's mission of helping to develop an informed debate on the impact of advances in medicine on people's lives inevitably implies concentrating on contemporary science. Our parent body, the Wellcome Trust, is a key funder of UK research in biomedicine: medical research itself and the basic sciences necessary for developing understanding of human illness.

The public understanding of science, even when restricted to this field, is still an enormous area. It was agreed, therefore, that to make a significant impact, the Centre should further focus its activity, including the funding of external applications, on genetics. Genetics was selected as our initial target, partly because of the Trust's investment in the area, but mostly because of its significance in the public arena.

The main target audiences of the Centre are:
- Secondary-school teachers and their students
- Primary health care professionals. Modern genetics is advancing at such a rate that the knowledge of most professionals educated more than five years earlier will be seriously out of date. We provide a variety of training for such groups.
- The non-specialist public. It is the general public, via the media and their representatives, who need basic information so that the necessary open debate can take place.

To date, we have funded fundamental research on knowledge and attitudes to genetics and genetics research among both children and adults, the production of several types of material to help teachers teach the science of genetics and various approaches to creating open discussion of the issues raised by increasing knowledge of genetics. Details of previous grants and guidelines for applicants are published in Wellcome Centre reports.[6]

The next step: collaboration

For many grant recipients, the main contact with a funding agency is at the beginning, when applicants submit research proposals, and, if successful, at the end, when reports are submitted on the outcome of the research.

There are other possibilities, however, whereby the funder provides resources in addition to money. In the case of the Wellcome Centre, these resources consist of some specialised knowledge, but more importantly, because we are part of The Wellcome Trust, we have access to a substantial proportion of the UK biomedical scientific scene. We have the power to open doors for our partners.

Of the £1500 million spent each year on medical research in the UK, the Wellcome Trust contributes about £300 million (about the same as the Medical Research Council). The Trust provides funding for part of the research of some 4000 scientists and is responsible for the salaries of about 3000 (figure 3). Grant applications are all peer reviewed and some 3500 scientists help us by acting as referees. Accordingly, we not only have access to the scientific scene, particularly in the UK, but an intimate knowledge of it. Scientists are busy people, but because so much medical research is funded by the Trust, many are prepared to help when they might not otherwise do so. However, we take care not to exploit this relationship. Research councils, government departments, broadcast and print media and others are also more readily prepared to collaborate because of the Trust's size in the funding arena.

Collaboration case studies

In the remainder of this paper, I describe three different examples of collaborations in which the Wellcome Centre has been involved, and the benefits obtained from them. I also raise some of the problems and pitfalls. Successful grant applicants always specify carefully what their objectives are, but if an active collaboration is involved we have found that great care is needed to ensure that both partners understand and share them. It is also true to say that, when the Trust is involved in such partnerships we too share 'ownership', which

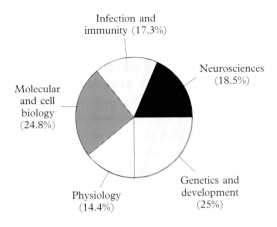

Figure 3. Proportion of Wellcome Trust funding for different areas of contemporary biomedical science (averages 1991–96)

brings with it trust and long-term commitment.

Not only do different partners bring different viewpoints or expertise, they also sharpen the discussion, which has its costs, too. Working in partnership can be exciting and productive. On the other hand, as there are more people to satisfy, it can take longer to agree shared objectives. Without such agreement, sharpening the debate can descend into unproductive squabbles, waste a lot of time or lead to compromises unsatisfactory to both sides. An obvious point here is that it is important to be clear not only about objectives, but also on relative responsibilities.

The first two examples of collaboration, Science Line and *The Gift*, were established following successful applications for funding. The third, with the Science Museum, was based on a long-standing relationship.

Science Line and ScienceNet

It is equally important that the public understand both the product and the process of science, and an experiment in the Netherlands—Science by Phone[7]—led the Wellcome Centre to fund a similar, though not identical, experiment in the UK. Not only does Science

Line offer members of the public the opportunity to obtain, over the telephone, authoritative answers to their scientific questions, it can also help to set callers on a journey: a journey of discovery in which, when appropriate, their questions are probed further in order to help them find the answers to their own questions by discussion with the scientists. ScienceNet is an electronic version for those with access to a computer and connection into the Internet. More than 35,000 members of the public have telephoned since the service was established in 1993, and there are currently some 6000 computer accesses each month. A selection of questions is given in table 1.

The full story of how Science Line was established has been described,[8] but the important point here is that it was established, with funding from the Wellcome Centre, after an application was encouraged from Broadcasting Support Services and Channel Four Television. Broadcasting Support Services is a not-for-profit organisation that provides telephone help lines for the general public. These include services after radio and television programmes and stand-alone services such as the AIDS helpline. Channel Four Television is a high-quality UK national network, funded by advertising.

Channel Four Television was an essential partner, as the first service provided by

Table 1. A selection of typical questions posed to Science Line by the public

- Can you tell me about a new cream for treating burns that was used in Bosnia?
- Why is sunrise so much shorter than sunset?
- How does DNA fingerprinting work?
- Do you know of any recent research into alleviating vertigo?
- What dangers exist from the lead piping in my house?
- If light is used to treat postnatal depression, this will reduce melatonin levels; will this in turn reduce prolactin and oxytocin levels, and is this a bad thing?
- Where in the brain are symbols processed?

Science Line was to answer queries from the public arising from programmes in their successful science series, *Equinox*. Planting and retaining the phone number in people's minds is the main problem in such services, and the Science Line phone number was trailed after the programmes.

Our launch service followed *The Real Jurassic Park*, a Channel Four Television programme examining the reality behind the fantasy of the feature film *Jurassic Park*.[9] There were 273,000 attempts to access our 20 telephone lines over the two days we remained open, and some 800 managed to get through. After this baptism of fire, things calmed down a bit with the other three programmes in the initial pilot. In each case, a team of specially recruited experts answered queries as the public phoned in after the programmes. As *Equinox* goes out on a Sunday evening, this says something for the altruism of our advisors, as we only pay their expenses. After the pilot series, a formal evaluation suggested that Science Line fulfilled a real need. Further funding was secured, and a daily service is provided from 13.00 to 19.00 Monday to Friday, while about once each month a specially recruited team is available after specific science programmes. An important feature is that the phone number is trailed after such programmes, thus maintaining the telephone number in the public eye.

A team of young scientists seeking to make a career in science communication has been recruited, and between them they cover most of mainstream science. Each works part time, so that we have two staff on duty at any one time. If a question is beyond the person answering the phone, they may wait until the relevant local 'expert' is on duty or they may call on one of the 1000 experienced scientists around the country who have been recruited to act as advisors for difficult problems. Collaboration does not stop there. The information services of the Wellcome Centre for Medical Science, the Science Museum, and the Natural History Museum also provide back-up help.

Broadcasting Support Services, which had appropriate technical facilities and expertise in providing information over the telephone, was crucial to the success of the service, but had no previous experience of providing a scientific service. This is where our links into the scientific community were helpful, and we and the information services above were an essential component in recruiting the national network of scientists advising the Science Line team.

The BBC now trails the telephone number after some of their programmes and the *Independent*, a quality national newspaper, runs a column called 'Technoquest' based on questions submitted to Science Line.

Once Science Line was established, we sought to develop it via Internet, and here another partner came on the scene: British Telecom in the guise of Campus 2000, now called CampusWorld. British Telecom provides the computer infrastructure, and Science Line scientists and advisors provide the scientific input. ScienceNet consists of a database of more than 1000 questions and answers covering the whole field of science, categorised into three levels according to age or knowledge. Answers are reviewed by an editorial board of school teachers and scientists to ensure accuracy and an appropriate level. Questions can be browsed or searched using key words and this service is available to anyone on the Internet.[10] Subscribers to CampusWorld, which include some 4000 schools in the UK, can also pose their own questions by e-mail and they receive a personalised answer by the same route. If appropriate, such questions and answers are then placed on the database.

It is obvious from the description above that running such a service is extremely expensive, and it is unlikely that any single organisation would be prepared to run one by itself for any length of time. Other funding partners—the Royal Society, the Office of Science and Technology (the source of government funding for science in the UK) and some of the research councils—have now come on board:

such a consortium is essential if the service is to continue.

Collaboration permits the provision of a national service at minimal, though not trivial, cost. Without each of the partners in the scheme, Science Line and ScienceNet simply would not exist. Inevitably, with such a large number of partners, difficulties arise because of different priorities accorded to Science Line or ScienceNet in different organisations. Moreover, while the objectives of the various partners overlap, they differ in detail. Accordingly, developing a successful consortium requires general acceptance that compromise is necessary for the good of the enterprise as a whole. Finally, while the large number of funders reduces the impact if one of them withdraws, it also reduces the visibility of individual funders who may thus find it less attractive to provide such funds.

A play about genetics

Some argue that the new genetics raises no new philosophical questions[11] but the public certainly thinks it does, and, as stressed previously, an important part of the Centre's mission is to help create an informed debate. This does not mean trying to reassure the public that there are no problems. Our aims are on the one hand to try to dispel unrealistic fears, such as those created by the film *Jurassic Park*, and on the other to damp down unrealistic optimism such as universal gene therapy for all manner of diseases which have a genetic basis. In the middle ground there remains a great deal to do.

A basic understanding of heredity is required to discuss the options for treatment of inherited disorders such as cystic fibrosis, Friedreich's ataxia or sickle-cell disease. There are issues about the main options currently available which need to be discussed by society at large. Our aim is to try to help ensure that decisions are taken in the light of an informed debate, not a knee-jerk reaction.

We felt it was particularly important to target 14- to 16-year-olds, as they and the next generation are the group whose lives will be affected in major ways by advances in genetics. Conventional approaches are unlikely to appeal to this audience, which is a notoriously difficult group to reach. It is nevertheless important that they learn to discuss such issues in an informed way. Science teachers generally feel uncomfortable with such discussions. Accordingly, we were on the look-out for a new approach when we were approached by Y-Touring, a theatre-in-education drama group with experience in dealing with subjects such as sex education and AIDS. Their proposal was for a competition for authors to write a play on genetics which the Y-Touring company would perform in schools, with an expectation that some 12–15,000 children would see it during the tour. Y-Touring had previously produced a short piece of participatory drama for us, entitled *Why Are We All Different?* This was designed for 8- to 10-year-olds and was a great success. This experience gave us confidence that they could actually deliver what they claimed, and referees' comments on their application were also positive. The proposal was accepted and the collaboration started.

The collaborative process

Y-Touring were able to identify promising writers and people from the world of theatre who could help us select a playwright, and provide continuing advice as the play developed. For our part, we were able to recruit geneticists as well as psychologists to ensure technical accuracy. Through another of our contacts, we also introduced the writers to patients suffering from inherited diseases, who gave of their time very generously indeed.

In due course, Nicola Baldwin's winning outline was turned into a play called *The Gift* (figure 4). This tells the story of one family over three generations struggling with the impact of an inherited disorder, Friedreich's ataxia. It not only describes in an unforgettable way some simple genetics, but also raises

many issues with which modern genetic research challenges society. An important feature of the project is a workshop which follows the performance of the play in both school and public performances. In the workshop, the issues raised by the play are extended and explored in a dramatic format. In schools, the actors retain the attitudes they portray in the play, but in the public performances we were able to recruit geneticists to take part. Although very positive in retrospect, some of the geneticists taking part were nervous about venturing into relatively unknown territory of public exposure.

Making it work

An essential ingredient for success was a deep commitment from both sides to make a success of the partnership. To begin with, Y-Touring felt our involvement indicated a lack of trust on our part, particularly our insistence on the involvement of a professional geneticist. To some extent they were correct. We were taking a risk venturing into areas of which we knew little. On the other hand, aside from our financial support, Y-Touring would have found it difficult if not impossible to find, let alone utilise, the scientific manpower we were able to harness by virtue of our standing in the medical research world. For our part, we could not have attracted our target audience or entered successfully into the world of the theatre. Both partners felt the collaboration was a great success, both sides bringing expertise and networks not available to the other, and our different perspectives led to a genuine cross-fertilisation of ideas during the development of the play. As we worked together, moreover, confidence grew on both sides and the end result delighted us all, so much so that the Centre's advisory committee funded a second tour that included the Edinburgh Arts Festival Fringe. For these performances we had the good fortune to be able to recruit geneticists from the Medical Research Council Human Genetics Unit in Edinburgh who were exceptionally helpful. Evidently they

Figure 4. A scene from a performance by Y-Touring of The Gift: *a drama piece about Friedriech's ataxia, funded by the Wellcome Centre*

found it a moving and enlightening experience too, as two of them wrote a glowing review in *Nature*.[12]

The success of *The Gift* has convinced us that drama is a powerful way of introducing the ethical and moral dimension to scientific issues and we have now commissioned another play from Nicola Baldwin and Y-Touring. It will deal with mental illness, the next focus of our activity in public understanding of science. Mental illness affects one in four people at some time in their lives and there is still a great deal of stigma associated with it. The Wellcome Trust invests a great deal in neuroscience, including psychology and psychiatry (figure 3), and, in addition to trying to help people to recognise the signs and to talk more openly, we wish to emphasise the importance of continuing research into mental illness. Our experience with *The Gift* leads us to expect that drama will be an effective way of handling these issues.

The Science Museum

The Wellcome Trust has a collaborative relationship with the Science Museum going back

more than 20 years. Sir Henry Wellcome's huge collection on medical history was loaned permanently to the Museum in the 1970s. These collections still constitute about 50 per cent of the Museum's entire archive even though many other parts of Sir Henry's acquisitions had been distributed elsewhere. Several thousand of the most significant artefacts have been on display in the Wellcome Galleries since the early 1980s, and the Trust recently supported the provision of high quality open storage for the rest off-site.

In 1996, the Governors of the Trust made by far their largest contribution to the public understanding of science, by providing the Science Museum with £16.5 million to add to the £23 million provided by the Heritage Lottery Fund towards a new 10,000 m² development to be known as the Wellcome Wing. In addition, the Wellcome Centre has provided funds for a travelling exhibition.

The Wellcome Wing

The Science Museum has long had ambitions to develop the west end of its site to provide space for contemporary science exhibitions, especially biomedical science in which the Museum is rather weak. An application to the Heritage Lottery Fund[13] was developed and the Governors of the Trust were asked whether they would provide partnership funding. The answer was positive. There were many reasons for this, one being the mutual respect in which each organisation held the other. This had been developed over years of managing the Wellcome Collection, when senior staff had come to know each other. The Trust had every confidence, on the basis of experience, that if funds were provided, they would be well used. A further factor was that two of the three subjects selected by the Museum for creating an exhibition on modern science were genetics and neuroscience—both subjects emphasised by the Trust in its funding portfolio. Even the third subject, information technology, is relevant, as it is crucially important in trying to make sense of the vast amount of information coming out of DNA sequencing and in modelling brain function. Another reason for the positive

response from the Trust was that we felt we could undertake a more effective public understanding of science campaign in collaboration with the Museum than by continuing to do so by ourselves. The Wellcome Building in Euston Road, which houses our exhibition *Science for Life*, is neither suitable for public access nor in an appropriate place to attract large numbers of visitors. The Science Museum has neither of these disadvantages, attracts nearly 20 times as many school children as come to *Science for Life*, and has world-class staff in all its many sections. The stunning design of the Wellcome Wing (figure 5) and the exciting exhibitions being developed, increase the attraction of collaboration still further. An established fruitful working relationship, respect for the senior staff involved and an overlap in terms of each of our goals were all important factors that led the Governors of the Trust to make this award.

The exact nature of the future partnership is now being explored. It is anticipated that, in addition to the provision of capital, part of the

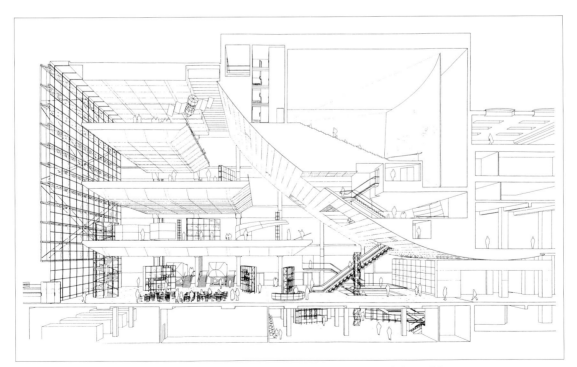

Figure 5. An artist's impression of the proposed Wellcome Wing at the Science Museum

Wellcome Trust's public understanding of science activity will be based at the Science Museum, once the Wellcome Wing is opened. What is gained by both is a pooling of resources and expertise. However, it has to be recognised that both lose some independence and will have to learn to work in unfamiliar ways. Furthermore, the Wellcome Centre activity will be on two sites, which will make our internal communications more difficult. However, in our view, the potential benefits far outweigh the risks inherent in such an enterprise and we are confident that together we will be able to achieve more than either could separately.

Travelling exhibition

A further collaboration, agreed in July 1996, is to develop a travelling exhibition together. It may be instructive to rehearse the reasons for this collaboration and some issues that have come to the fore already.

Over the past few years the Science Museum has developed a series of small, travelling exhibitions on contemporary science and technology.[14] In 1995, the Museum was seeking financial support for exhibitions that presented a topic in contemporary biomedical science. The Wellcome Centre has developed travelling exhibitions itself, so why collaborate? The answer is simple. While undoubtedly valuable, travelling exhibitions are expensive to develop and run. The Science Museum wished to develop exhibitions in the biomedical field and our missions thus overlapped. It therefore seemed sensible to pool resources, provided we could agree on details of the arrangement.

A key point that gave the Centre's Advisory Committee some difficulty was the issue of editorial independence. The Wellcome Trust prides itself on its own independence as it depends on neither the collecting box nor government for its funds and it has the backing of a large proportion of the medical research community. (It was the Trust, after all, that stepped in and funded a national survey

of sexual lifestyles in 1993 after the government of the day vetoed the plan of its own Research Councils who considered this an important study and had set aside funds to finance it.) For such a body to yield editorial control to another requires confidence that that trust will not be betrayed. On the other hand, clearly the Museum has to retain editorial independence from sponsors if its exhibitions are to be given credence by a sceptical public, and therein lies a potential for conflict.

However, I believe there is a fundamental difference between the normal type of sponsorship and that from the Wellcome Centre. We have no 'corporate' or 'marketing' agenda, as we have no commercial interest in the work we support. This means that collaborative relationships are different from those usually involved with an industrial partner. What we are trying to develop is a partnership in which, while the Museum obtains the funds it needs and will provide the greater proportion of expertise, we will contribute both staff time and access to our network of Trust-funded scientists. The perspective from those doing the research, those at the cutting edge, is not necessarily the same as that filtered by communicators of science and this refreshing, direct contact is a powerful asset that we bring to collaboration.

Potential pitfalls remain. It is conceivable that problems could arise similar to those encountered by the National Museum of American History (part of the Smithsonian) in their exhibition *Science in American Life*, sponsored by the American Chemical Society. While some scientists found the exhibition true to life, others, including the sponsors, felt that the exhibition portrayed scientists in an unduly negative light[15] and major tensions arose. There is no simple way of avoiding such differences of opinion except by detailed discussion and mutual education. In our situation, both the Science Museum and the Wellcome Centre agree that an honest description of science and technology, warts and all, is more likely to win hearts and minds than the triumphalist mode so often adopted.

Conclusions

The Wellcome Centre provides money for initiatives in the public understanding of contemporary medical science because it feels that an informed public should be involved in the important decisions that need to be made about the use of medical technologies, now and in the future. A positive climate of opinion is also more likely to allow medical research to flourish.

In this paper, I have stressed the value we place on collaboration. Providing collaborating partners share a similar vision, there are economies of scale, sharing of expertise and access to different networks otherwise inaccessible or at least much more difficult to reach. It can also be a lot of fun.

There are some points to bear in mind in order to help ensure a successful collaboration. One obvious but sometimes forgotten essential, is to be clear about the overall objectives of the collaboration. Objectives will naturally be set down in writing for a grant application, but even then, different perceptions can lead to difficulties. Good communication is the key and extra special efforts are required to ensure that this is effective. The question of editorial control is not a trivial one. Whether one funds an exhibition or a play, one puts oneself at the mercy of someone else who also has an agenda. There is no single truth, and yielding editorial control to another is an act of trust which will happen only once trust has been established. It is also essential to develop a clear project plan so that everyone involved knows who should be doing what and when, in order to ensure that all essential work is completed on time and that there is no unnecessary duplication. Effective monitoring is also required. Last, but perhaps most important, effective collaborations are difficult to develop without trust: trust in each other's competence and trust in each other's integrity. And trust takes time to develop.

I have tried to show how partnership is inherent to the Wellcome Trust's whole way of thinking. Science is increasingly an interdisciplinary activity, in which each of the partners brings their own special skill or experience to bear on a problem. The combination sometimes sparks unexpected novel solutions. So, too, with the two-way communication between the scientist and the public. The public's desire to know about science is ever changing. Different publics and different issues require different approaches and different media, and no single institution can provide all the skills necessary. Outside specialists, whether in drama, exhibition development, computer graphics or telephone help lines, will already have the necessary skills and be at the forefront of their subject. Ignoring such opportunities or trying to develop special skills in-house when required, condemns one to stagnation or being behind the times. In our experience, collaboration has been exciting and productive and has led to more innovative outcomes than we could possibly have achieved alone. I recommend it.

Notes and references

1 The bursaries are administered by the Association of British Science Writers and a total of six bursaries are available annually: two each for the MSc in Science Communication at Imperial College of Science, Medicine and Technology, London, two for the course in Journalism at the City University, London, and two at Techniquest Science Centre, Cardiff.

2 Rhodes-James, R, *Henry Wellcome* (London: Hodder and Stoughton, 1994)

3 Markby, Steward and Wadesons, *A Copy of the Will of Sir H S Wellcome, and a Memorandum for the Guidance of his Trustees* (1936)

4 The Medical Research Council is the UK equivalent of the National Institutes of Health (NIH) in the US, but with a budget of £380 million instead of the NIH's £7.4 billion.

5 Venues for *Genes Are Us* have included Euston Station, London; Gyle Shopping Centre, Edinburgh; Metro Centre, Newcastle; Galleries Shopping Centre, Bristol; Ideal Home Exhibition, London; Eagle Shopping Centre, Derby; Pallasades Shopping Centre, Birmingham; Gare du Midi, Brussels.

6 Further information is available on the Trust's World Wide Web pages (http://www.wellcome.ac.uk).

7 van den Broeke, M, 'What the public wants to know', *Science Communication in Europe, a Report of a Ciba Foundation Discussion Meeting* (London: Ciba Foundation, 12–13 June 1991), pp84–87

8 Holland, S, 'A line to science. The development of Science Line', *The Wellcome Trust Review*, 4 (1995), pp60–63

9 *Jurassic Park* (directed by Steven Spielberg and written by David Koepp and Michael Crichton) was released in 1993 by Universal Pictures. In the film, dinosaurs were brought back to life through the cloning of DNA extracted from blood-sucking insects trapped in amber. The Channel Four Television programme, *The Real Jurassic Park*, produced by Windfall Films, was screened on 18 July, 26 September 1993 and 2 August 1994.

10 http://www.campus.bt.com/CampusWorld/pub/ScienceNet/

11 Wolpert, L, 'Is science dangerous?', *W H Smith Contemporary Papers*, 15 (1996)

12 Wright A F and Boyd A C, 'Choosing genes', *Nature*, 383 (1996), p312

13 The British government established a national lottery in 1994. Prizes are substantial, and to the despair of those interested in public understanding of statistics or of risk assessment, the public has spent freely: so far nearly one billion pounds has been raised for charitable causes. The funds are divided into several sections, one being the Heritage Lottery Fund. This is confined to public-sector agencies, charities and voluntary organisations. Institutions may seek funding from the fund, provided they can make a convincing case for heritage merit and public benefit. All applicants are expected to provide a proportion of non-government funding towards the proposed project.

14 Ward, L, this volume, pp83–90

15 Macilwain, C, 'Smithsonian heeds physicists' complaints', *Nature*, 374 (1995), p207; Macilwain, C, 'Now chemists hit at Smithsonian "anti-science" exhibit', *Nature*, 374 (1995), p752; Molella, A, this volume, pp131–37

Developing partnerships for the display of contemporary science and technology

Gillian Thomas

Partnerships with industry provide important resources for new exhibitions. This is an essential guide to the art of building and maintaining good relationships.

All organisations need supporters, all organisations need resources. This is especially important when planning and developing exhibitions which feature contemporary science and technology. A partnership strategy aims to identify those resources of a potential partner that can be of assistance to the organisation and to develop a relationship that is supportive to both. This article looks at the potential types of resources, both of the organisation seeking to develop a partnership and of the possible partner, and explores how a mutually beneficial strategy can be developed. The *Challenge of Materials* gallery at the Science Museum, due to open in May 1997, is considered as a case study of a partnership strategy.

A partner in a commercial organisation shares in both the risks and the profits. The same is true for a science centre or museum aspiring to develop a partnership scheme. Choices about who the partners are and how the relationship is structured will influence whether the risks or the advantages predominate.

Several key sectors can be considered; each has its own differentiating characteristics:
- The main aim of the *corporation* is to make money. Positioning the corporation in the perception of the general public is important. Social aims may be part of this positioning strategy, as will be the welfare and opinion of its workforce.
- A *trade organisation* will be interested in the overall perception of the industry, as opposed to that of a particular company, but the larger companies will probably have a great impact on deciding policy. The trade organisation may be the major forum for industry debate.
- *Foundations and trusts* each have specific purposes, from which they cannot deviate by law. Personal contacts within these organisations are often of importance in interpreting how the foundation or trust's published aims will be applied.
- *Private donors* each have their individual interests, and personal contact is fundamental.
- Whatever the tier of *government funding*, the issues are political. Because many museums and science centres are dependent on government funding for a substantial part of their resources, they themselves could be considered a part of a governmental strategy in areas such as the promotion of industry, or education and training.
- *Employee bodies* have not been traditionally considered as potential partners for organisations. However, in a society that is reviewing its attitudes to training and lifelong learning, these could become interesting partners.

Corporations and trade organisations are particularly interesting, as, on the surface, they would seem to have little reason for altruism to lead them to substantial financing of a not-for-profit organisation. However, as it is in these sectors that applications of the latest scientific and technological research are developed, they could form valuable partners for museums and science centres. It is thus worth examining this sector in more detail.

Corporations have a varied range of resources which could potentially be of interest. These include: financial assistance;

staff on secondment; know-how and advice; access to the latest technological development; a workforce who are potentially visitors; press and promotions programmes; publications; and networks of contacts. A corporation will usually have a long-term strategy, so that its potential as a partner can be structured over a number of years. It is important that this long-term strategy is understood by the organisation seeking support. Before a plan of action is developed and the corporation is approached, the strategy must be taken into account to ensure a close match between the project's aims and the business interests of the potential partner.

Trade organisations will have a particular interest in the public profile of the industry and trends in public opinion related to issues on which the industry could be seen to have a significant impact. They will be interested in the broad spread of the industry, and its present and future general profitability.

What do museums and science centres have to offer?

At the first glance, it may seem as if the industry has all the advantages stacked on its side and that it would only enter into a partnership in fulfilment of a social policy. However, a museum or science centre has a number of advantages to offer industry as a viable partner.

A not-for-profit organisation has privileged access to the public. This covers both the general public, specialised interest groups and educational visitors. The numbers concerned are substantial: the Science Museum alone received over 1.6 million visitors in 1995, of whom over 300,000 were in prearranged school groups. The museums and science centres within the European Collaborative for Science, Industry and Technology Exhibitions organisation in Europe (ECSITE) welcomed over 26 million visitors in 1996.[1] These individuals have, largely, chosen to visit. They are therefore potentially in a receptive frame of mind.

As part of the development, or as a separate programme, a museum may be carrying out research of interest to the partner in the field of the public understanding of science and technology. The results of such research can also be of use to generate well-positioned press coverage.

Information presented in a museum or science centre is seen by the public as impartial. It is not perceived as publicity for the government or primarily as promotion for the industries concerned. There is an expectation on the behalf of the public that information related to an industry will be correct. This is of inestimable value to the industries, and, for the position to be maintained, it is essential that the museum or science centre at all times retains the right of editorial control. This does not exclude the museum from working closely with the industrial partners and ensuring their position is fairly represented. But it does ensure that the exhibition or activity is not simply an advertisement for the industry and that the public will continue to consider the content of exhibitions as factual.

Science museums and centres have an unusually high incidence of specialist visitors. A survey at the Science Museum on an autumn weekday revealed that, excluding prearranged school groups, 20 per cent of the remaining visitors were single men with a specialised interest in science or technology. Any display related to their specialism is thus likely to attract their attention and the positioning of the particular industry will be reinforced.

Similarly, once a specific exhibition has been created, it can be used by an industry to present its products to particular audiences, such as commercial fund managers, specialised press, sector professionals and their professional organisations, using a series of tailored events. Displays of particular products can also be incorporated into these events.

A museum can create events which will attract press coverage of an entirely different nature and exposure compared with promotional events organised solely by the corporation. The general press and television channels will regard exhibitions and events, within and organised by the museum, in a completely different light than those organised by a company for purely marketing purposes.

An exhibition can be developed with the needs of corporate hire in mind. This makes it much easier for the company to exploit the resource subsequently and the spaces so designed can generally be adapted for other group use, such as schools. Thought given at an early stage of development as to how these requirements can be integrated will substantially affect the use that a partner can make of the facilities.

At another level, it should be remembered that the museum project may offer a degree of interest and excitement to executives within an organisation. The influence of personal pleasure in a decision as to whether to proceed with a partnership is perhaps as important within a corporation as in a foundation.

Finally, many larger organisations actively seek to assist the personal development of staff. The museum can offer a range of opportunities that can be integrated into such schemes. These include free entry for staff and families, staff parties for special occasions and training opportunities.

Establishing a partnership

Any partnership takes time to develop and must be pursued at different levels over a period of time. Once a potential partner has been identified and a possible line of approach for a project imagined, initial contacts need to be made while the project is still in the state of development. If a substantial collaboration is envisaged, these contacts should be made while the research on content is still going on and the exhibition form not entirely finalised.

While this may seem dangerous, as it could be argued that the organisation runs the risk of skewing the content to please potential partners, it does mean that advantage can be taken of the partners' technical expertise and knowledge of latest developments, at a stage when the content is fluid, as opposed to

endeavouring to assimilate unwelcome additions at a later stage. The partners can also become involved in some of the more pleasurable aspects of exhibition design and development, by seeing how the design develops.

The first steps in establishing a partnership are the most difficult and the skills of a professional are needed. In any large organisation, contacts at different levels are essential. A board-level contact will need to be supported by assiduous work at local level, to ensure that all relevant officers within the corporation are aware of the approach. Relying only on top-level contacts can be dangerous, as few employees enjoy simply carrying out someone's instructions. If, on the other hand, a chief executive officer or board member asks the corporate affairs department for their opinion on a project and they are already well informed, they appear to good advantage. In contrast, if a project has worked its way up to board-level from the local level without any high-level lobbying and support, it risks being rejected simply because no one has heard of it and board members have other favourite schemes.

What exactly should be presented at these early stages varies. Generally a short four- to five-page document is sufficient, clearly outlining the project and giving its background in the context of the host organisation. Opinions vary in different countries as to at the stage at which the financial implications should be mentioned. It is helpful if the overall range of costs is identified early on, but a structured package will take time to develop. Too detailed a package at an early stage can give the impression that a project is fixed and costings may, in any event, not be entirely accurate.

Structuring the deal

It is unlikely that any partnership would include all these potential advantages outlined above. At the beginning of the partnership, a package of advantages must be agreed and incorporated into a contract. This has benefits for both parties:

- The museum is clear about what it is offering and accepting. It is then less easily pressurised into giving further advantages, either by the corporation or by team members on the project.
- The corporation has a clear statement about what has been agreed. This is of particular importance in a large organisation, as the terms of any agreement can become obscure within the different parts of the organisation and over the life of the project, as staff change and different individuals become involved. A clear written agreement forms a point of reference.

Within a large project, a complex structure may evolve within the different packages envisaged for different partners. This is one of the advantages of working with a trade organisation. They can then act effectively as the sole negotiator for the industry, which reduces the problems associated with a large number of potentially competing partners. Similarly, through a trade organisation, it is possible to incorporate competitors more easily. If only one main partner is involved, other competitors in the same sector will be less likely to be integrated. A mechanism needs to be agreed within the package to enable the integration of subsequent partners, probably offering less financial support, from different industrial sectors, but who also require recognition.

Elements to be incorporated into the agreement include:

- a clear description of what the project is, its size, budget, scope, audience, duration, the development and construction programme, and research to be undertaken.
- a list of potential benefits to the partner, including how and on what the partner's name will appear.
- a description of what the host organisation will be putting into the deal, including staff time, financial support, event organisation, educational resources, facilities for corporate hire and public-relations support.

Corporations are interested in how their image will be maintained throughout the life of the exhibition or programme, not just at the outset of the project. For this reason, it may be possible to encourage them to spread their support over a number of years, to ensure that the contemporary nature of the project is maintained. The integration of substantial events or updating programmes may, contrary to expectations, increase the potential partners' interest in a project.

Developing a partnership is a long-term job: signing the deal and getting the first payment is important, but a partner has then been acquired who will require services throughout the life of the partnership. There can be many areas of concern. Is the museum delivering what it promised? If there is a financial or staff commitment on the part of the museum, is this being met? Is the project, now completed, living up to its expectations? If not, has any money been kept for alterations? Is the partner's corporate logo correctly displayed on all subsequent publications? Is there continuous and favourable press comment?

As the number of partners of a museum increase, this servicing will take up an increasing amount of time. As staff change and the pressures of current projects grow, it is tempting to allow this essential following through to be insufficiently covered. This, however, is dangerous. A satisfied partner can bring in many others and may also increase their participation, whereas a dissatisfied one may inhibit others from joining.

Challenge of Materials

Challenge of Materials (figure 1) is an exhibition about materials, what they are, how they are made, how they are used and what happens to them at the end of their useful life. This new exhibition opens in May 1997 and includes a ten-year programme of events, updates and temporary exhibitions. It replaces three galleries, some 20 years old, devoted to steel, plastics and glass.

Curators, project members and sponsorship staff collaborated to develop a concept based on the notion of a 'challenge'. All three galleries were to be replaced with a single exhibition, focusing on the materials industry and materials science, as one of the newer areas of science and technology currently under-represented in the Science Museum. It was felt that this combined approach would offer a more balanced view of a substantial industry and would be of greater interest to the general public.

developed to include other aspects of the materials industry, including plastics, glass, aggregates, aluminium and rubber. In each case, where appropriate, the trade industry has been involved, although it has not always become the eventual contracted partner. From the beginning, the partnership packages have been structured to integrate the different

Figure 1. *A computer-generated image of the glass bridge for the* Challenge of Materials *gallery. The bridge, which will be supported by steel wires measuring only 1.6 mm in diameter, will form an interactive exhibit with lights and sounds that respond to the changing load conditions.*

From the three main industries with a previous presence in this space, a major sponsor was sought who would effectively lead a materials industry partnership. It was decided that the UK Steel Industry was the most appropriate, given the size of the industry, its long history in the UK and its involvement in every area of manufacturing and construction.

As an approach was sought to the whole of the steel industry, the appropriate body was the trade association, BISPA, for which the major supporter is British Steel.

Once the main potential partner had been identified, subsequent proposals were

industries and the varied levels of support. This both reassures the potential partners and reduces the individual negotiations.

Because the programme has a ten-year life span, its final success can only be evaluated at the end of the ten-year period. Some clear characteristics have, however, emerged during development.

First, it has shown that the support that active partners can offer during the detailed development of a project is invaluable. As a supporter of the project, the industry feels a sense of ownership and is thus endeavouring to give every possible assistance, in addition to financial support. This includes supplying materials, technical advice, access to film and other archive material, visits to sites for research and staff secondment. As opening approaches, a coherent press strategy is being developed with all the partners, making use of their extensive resources.

Second, the importance of the ten-year programme, as a guarantee to the partners that the exhibition will not lose its contemporary nature and interest to the public, has been crucial. Over the ten-year period, programmes of events and drama will be supplemented by two update areas which will change every six months, and a renewable, high-profile core display that changes every 18 months and will enable the exhibition to be regularly re-launched.

Third, there has been tremendous interest from industry in showing its latest products, as part of the *Technology Showcase* series of small temporary exhibitions integrated in the space. One difficulty that has arisen is that many of the latest products which capture the interest of the press and the public are often of only fringe importance for industry. Areas such as nanotechnology, electrorheological liquids and shape-memory alloys may fascinate the public and must be included, but they may not currently be of significant economic impact and do not necessarily correspond to what the industry sees as new products likely to find substantial application. This can, however, be put to advantage. What better place to show a new product, for which a full range of applications has not yet been devised and with the corporation currently seeking interested partners?

Finally, a balance is needed in the exhibition between the contemporary, which will need regular updating, and some underlying themes which should endure for the life of the exhibition. Here it is important to distinguish between physical renewal, of exhibits which remain intellectually valid but none the less will need replacing through being worn out by use, and those exhibits for which the content is out of date. Too great an emphasis on contemporary science and technology will tend to mean that the exhibition will date rapidly and be difficult to maintain. One of the main requests from the public is for increased interactivity. These types of exhibits, however, take the longest to develop and the have the highest maintenance costs. It is advisable that the majority of exhibits which will require updating within a minimum of a year should not be of an interactive nature, as the development investment will be high and insufficiently exploited.

Conclusion

One of the aims behind the founding of the Science Museum, nearly 150 years ago, was to ensure that the general public was aware of the latest products being produced by industry, as a means of encouraging the industrial development of the UK. This educational aim is still part of our mission today. As the technological capability of the workforce becomes of increasing importance, there are opportunities for science centres and museums to work with industry to raise the levels of awareness of the population, to present issues that relate to people's lives, and to illustrate the diverse implications that industry has on the future prosperity of the nation.

The types of collaboration are many and varied—the scope enormous. In all cases, it is essential to remember one thing: this is a partnership, each has something to offer. It is up us to make the best of what we have to share.

Notes and references

1   Staveloz, W, Executive Director, European Collaborative for Science, Industry and Technology Exhibitions, Brussels, personal correspondence (1997)

Stormy weather: *Science in American Life* and the changing climate for technology museums[1]

Arthur Molella

The 'science wars' are changing the ideological environment and museums are feeling the heat.

This paper forms a kind of cautionary tale, North American style. I cannot say that these lessons are applicable to other institutions or countries, so make of them what you will. Let me begin with a brief recapitulation of funding realities in the US today. Public museums like the Smithsonian no longer receive sufficient public funds to mount large, or even medium-sized exhibits. In fact, for all practical purposes, the Smithsonian Institution is fully dependent on privately raised dollars—most often from corporate sponsors—for all major initiatives.

Our financial situation directly affects the tone and style of collaboration. It is a fact of modern life that sponsors expect a positive view of their product lines, especially in industrial museums. These expectations are sometimes tacitly expressed, but these days they are increasingly explicit. Of course, it is important to present social issues and controversy in critical perspective in science museums and centres. But, in our present era of heavy dependence on corporate sponsorship, presenting contemporary science in full social dress can be problematic.

I call my story 'stormy weather' (with apologies to Lena Horne), because it commences on a snowy Thursday evening in early February 1995 at the National Museum of American History. The public had gone home, leaving the halls of the museum deserted and dark—except for the entrance to the *Science in American Life* exhibition, where I stood before a tribunal of elite American physicists. Numbering among them two Nobel prize winners and the highest officials of the American Physical Society (APS), they had gathered from around the United States to put the universe back in order. The gods of science had been mocked and someone had to answer for it. After polite greetings and small talk, I began to lead the group on a tour of the exhibit for which I had served as chief curator. Before I had uttered more than a few words of welcome, a brusque voice from the visiting delegation interrupted, rudely shattering the mood: 'Do you have any idea why scientists are angry about this exhibit?' I won't reveal my reply just yet, but only say it did not get any more pleasant after that.

Indeed, the officials of both the APS and the exhibition's underwriters, the American Chemical Society (ACS), were sufficiently annoyed to launch a strident media campaign against *Science in American Life* and to demand revisions in the presentation—a situation virtually unprecedented at the Smithsonian. To the exhibition's curators, the attacks by the ACS seemed especially inappropriate in light of the Society's prominent role on the exhibition's advisory committee, and the four years of collaboration that had gone into the planning of the exhibition. Spirited, sometimes heated, debates marked the advisory process, yet the ACS had ultimately endorsed *Science in American Life*. According to the ACS, it was the hearing given to the prior complaints of the APS that prompted their post-opening initiative against the exhibition.[2]

Masterminding the campaign was Robert Park, a University of Maryland physicist and head of public relations for the APS. It was Park who introduced God into the equation, accusing us of portraying science as the 'God that failed'—an ironic take on a title of Arthur Koestler's.[3] Although criticism of the exhibition came almost exclusively from these senior scientists, the societies proved a formidable special interest group.[4] And the Smithsonian was vulnerable. It was still reeling from the debacle of the *Enola Gay* exhibition, the National Air and Space Museum's ill-fated presentation of the B-29 that dropped the atomic bomb on Hiroshima. (In fact, just a few days before, the Smithsonian had officially cancelled the original *Enola Gay* exhibition.) Still struggling to limit political damage from that debacle, the Smithsonian Secretary—voicing his own worries that *Science in American Life* 'baited' scientists in the prevailing politically charged climate—ordered changes to the exhibition, not nearly so extreme as the senior scientists wanted, but changes nonetheless.[5]

What had we done to enmesh ourselves in the so-called 'science wars'—the recent assault of a surprisingly large number of scientists on science-studies scholars, primarily in academic institutions but now extending into museums? What does our experience suggest about the current situation of technology museums in the US?

There was a time not long ago when technology museums seemed immune to controversy. At the Smithsonian, the National Air and Space Museum was held sacrosanct, and technology exhibits were hardly ever criticised at the National Museum of American History. Our problem lay rather in getting exhibitions noticed by the press, and in engaging both journalists and the general public in thoughtful debate about science and technology. We were vastly disappointed, for example, when our *Engines of Change* exhibition on the American industrial revolution elicited only the most superficial reportage of the 'gee-whizz' variety— this, after we had worked hard to depart from the traditional presentation of machine as icon to address production, labour, consumption and other social issues.

What was it about *Science in American Life* that abruptly broke the pattern? What we attempted to do in the broadest terms was to implement in the realm of museum exhibitions a now-standard injunction to view science and technology, not in traditional ways as isolated phenomena, but in the context of societal changes and values. The exhibition's premise is that scientific research and science-based technology have emerged as the most powerful agents of change in America over the past 125 years, and that science has grown into a complex enterprise interwoven with all aspects of American culture.

Using about two dozen case studies from 1876 to the present, *Science in American Life* explores critical intersections of science, technology and society, including: the founding of a pioneering chemical laboratory in an American university; the uses of experimental psychology and intelligence testing; the mobilisation of science for the Second World War and the Manhattan Project; the attempt to control and improve upon nature through new materials and processes, from plastics, to nuclear power, to the birth control pill; and the new frontiers of biotechnology. Let me editorialise a little: *Science in American Life* is a science and society exhibition—a rare bird, to be sure—but its approach is entirely traditional. Its story is mainstream, the sort of narrative you would commonly find in standard texts on the history of science in America.[6]

A spacious hands-on centre at the beginning of the exhibition helps visitors understand how science works through learning by doing, in such areas as DNA, water testing, plastics recycling, radioactivity in common objects and so on. The centre has been exceedingly popular, receiving well over 500,000 visitors, mostly children, since opening. The sight of all those engaged children failed to soften the hearts of our scientific critics, however.

Now I'll tell you how I answered that physicist who asked if I knew why they were angry. I told him it was because the exhibition ran counter to many scientists' expectations. Traditional museum and science-centre shows make science itself the central story. Rarely, if ever, are social implications the subject. Dismayed by our unfamiliar social-history approach, our critics accused us of souring public attitudes by over-emphasising the problematic aspects of science, and slighting the spiritual dimensions of pure science.

In response to these charges, the Smithsonian commissioned an impartial evaluation of the exhibition's visitors. This assessment, conducted by the Office of Institutional Studies—which is independent of the National Museum of American History—demonstrated conclusively that visitors came away from *Science in American Life* with overwhelmingly positive feelings about science and technology.[7] Although these results took some wind out of the sails of our critics, they were not mollified. They particularly criticised our deliberate blurring of the boundaries between science and technology, our depiction, for instance, of many scientists operating in applied settings, rather than as pure discoverers. Maintaining a self-image as truth-seekers, they viewed this as tainting science. Sequestered in a laboratory, how could a research scientist cause problems? If there was blame to assign, it was to the appliers of knowledge, not to the truth-seekers. What worried the scientists most of all, however, was our entangling of science, not only with technology, but with society and culture in general: enmeshing it in real life, in other words. Any approach that failed to portray science as a pure endeavour, innocent of any societal involvement, remained suspect in their eyes—and dangerous.

The curators of *Science in American Life* were confused. Just four years earlier, a very similar exhibition at our museum, *Information Age*, came in for a very different sort of criticism from the media, and even from top Smithsonian management. The adverse criticism came from a direction 180 degrees removed from the guns aimed at *Science in American Life*. When *Information Age* opened in May 1990, the Washington press, both liberal and conservative, skewered it for being a tool of 'Big Brother'—for unabashedly promoting the products of the 20 or so computer corporations that paid for it. Most humiliating of all, at the opening press briefing, Smithsonian Secretary Robert McC Adams surprised everyone by attacking his own institution's

exhibition for its lack of critical cultural perspective.[8] Even though the high-tech firms that sponsored the exhibition tried to console the curators, what were the staff of the museum to think after this public rebuke from their highest official?

Paradoxically, in retrospect, there are more similarities than differences between *Information Age* and *Science in American Life*. This is not surprising, as many of the same people worked on both exhibitions and in the same institutional milieu. Both exhibitions cover the modern era, work with roughly the same periods, explore cultural implications (criticisms of the Secretary notwithstanding), and examine current issues of technology and society, some of them problematic (such as the privacy issues raised by the automation of Federal Bureau of Investigation's fingerprint files). In short, messages and exhibition styles were not notably different.

It is now clear to me that what had changed was the ideological environment. The political and ideological ground had shifted at some point between the openings of the two exhibitions, between May 1990 and April 1994. I'm convinced that if *Science in American Life* had opened during the 1980s or even early 1990s, it would have been spared most of the controversy or, just as likely, suffered attacks from the left. The story was very different in 1994.

In the interim, science and technology had joined the culture wars in the US. It soon became obvious that the attack on *Science in American Life* was just a skirmish in the larger 'science war' waged by scientists on science and technology studies—in books like Paul Gross and Norman Levitt's now notorious *Higher Superstition*,[9] and in journals like *Social Text*, where physicist Alan Sokal planted a phoney postmodernist piece on 'quantum gravity', while simultaneously exposing his hoax in another journal.[10]

My first inklings of what was really going on occurred during an after-dinner speech celebrating the opening of *Science in American Life*, in which a high official of the ACS warned of a growing anti-science movement in the US, driven, he alleged, by current trends in university humanities departments associated with the academic left postmodernism, relativism and deconstruction. The ACS, he vowed, would combat the foes of science and reason.[11]

He refrained that evening from explicitly attacking the very exhibition his society had sponsored; but the aforementioned Robert Park did not hesitate to attack the exhibition afterwards, invoking the same rhetoric against irrationalism, postmodern relativism and leftist ideology.[12] I find this odd: there is not a postmodernist sentiment in *Science in American Life*. It was revealing of ideological biases that Park's attacks on *Science in American Life* incited a number of scientists to denounce an exhibition they admitted never having seen. It was obvious that the exhibition was serving as a pretext.

In my opinion, the reasons for these scientific reactions are not mysterious: a widespread—and, I believe, genuinely felt—perception of lost prestige and power stimulated by funding cuts imposed after the end of the Cold War, the first budget cuts imposed on the American scientific community in decades—when, for the first time, science and science-based technology could no longer count on the unlimited and unquestioned patronage of the Defense Department. (I know there are similar developments in post-Cold-War Europe.) The Congressional cancellation of the Superconducting Super Collider in September 1993 became a symbol of the new fiscal realities. (It did not help the cause of *Science in American Life* that it featured a case study on the funding and siting of the Super Collider, figure 1.)

Contributing to this siege mentality were some high-profile charges of fraud and misconduct against scientists that had come from within the scientific community and that had attracted media attention and Congressional scrutiny. Intemperate Congressional rhetoric fanned the flames of culture wars. The recent highly publicised case of misconduct involving the now-exonerated Nobel laureate David Baltimore had thoroughly shaken scientists'

sense of status and security. For the first time in decades, scientists found their autonomy and the privilege of self-policing challenged by public officials. For the first time, they had to justify their funding and their actions to government funders and the public.[13]

While, in reality, science and technology remained generously funded by any standard, a malaise has beset the US scientific community. Feeling misunderstood and unappreciated, leading scientists are lashing out at public ignorance, and what they see as a deep-seated public scepticism over the value of scientific knowledge and the technology it spawns. Despite recent polling data collected for the National Science Foundation that shows no ebbing at all of a very high level of public esteem for science and technology, scientists have continued to bemoan their situation.[14] Seeking someone to blame for their perceived predicament, they have denounced popular hucksters and charlatans like Uri Geller, but also relatively obscure (and harmless) academics in Science and Technology Studies departments—all those postmodernist critics.

Within months of the opening of the *Information Age* exhibition, the culture wars began to have an impact on public museums. The Smithsonian's *West as America* art exhibition suddenly found itself under attack for 'political correctness', which led to Congressional scrutiny of other Smithsonian projects, including its celebration of the Columbus quincentennial.[15] All of this was aggravated, of course, by the Republican sweep in Congress in 1994 and the ensuing efforts to tighten controls on public cultural agencies, leading, among other things, to demands for the termination of the national endowments for the arts and humanities. The controversy over the *Enola Gay* crystallised the cultural debates as a war between scholars and ordinary patriotic Americans.

Scientific critics of *Science in American Life* saw an opportunity—our case study on the Manhattan Project—and immediately tried to link our exhibition with the *Enola Gay* and

Figure 1. Case studies on the 'Superconducting Super Collider' (left) and on the ozone hole, depicting Susan Solomon's Antarctic expedition that confirmed the role of chlorofluorocarbons in generating the 'hole in the sky'

with leftist academic scholarship. Because of the current media fascination with these issues, this gave their story a modicum of play, but not nearly as much as our critics wanted. Fortunately, their campaign never took off like the *Enola Gay* controversy. But they were able to exert enough pressure to force a Smithsonian response in the form of modest revisions of the exhibition.

That the ACS approached the Smithsonian with one set of expectations while the Museum had another contributed greatly to the difficulties of *Science in American Life*. But changing times inflamed the conflict and the public controversy. I can say that, five years ago, we had no inkling of the science wars and how they would affect science and technology museums. The abrupt shift in the winds between *Information Age* and *Science in American Life* shows just how quickly the atmosphere can change. This makes me wary of predicting what technology exhibitions will look like and how they will fare even two years from now. What we do know, however, is that science

and technology exhibitions are no longer sheltered from the elements. Private funding— from individual donors, corporations and professional organisations like the ACS—is increasingly becoming a necessity for the Smithsonian and other public museums in the US. So long as we depend on these sources we will be vulnerable to outside pressures, which inevitably worsen in stressful times. So, at least for the near future, before we venture too far outdoors, we'd better check the weather.

I would not want to end on a pessimistic note. The Smithsonian has indeed had many successful experiences with private donors, both individual and corporate. The Lemelson Center, which I describe in another paper,[16] is a stellar example of collaboration with a major individual donor. What we have learned from that experience is that agreement up front on common goals can ease the path of collaborations. In fact, we look forward to more collaborations of that type at the Smithsonian and to smoother sailing ahead.

Notes and references

1 A more extensive version of this paper is due to be published: Post, R and Molella, A, 'The call of stories at the Smithsonian Institution: history of technology and science in crisis', *ICON*, 3 (1997), in press.

2 American Chemical Society chairman Paul Walter's letter to Smithsonian Secretary I M Heyman was reported in the science press, in Macilwain, C, 'Now chemists hit at Smithsonian's "anti-science exhibit"', *Nature*, 374 (1995), p752.

3 Robert Park edits a weekly Internet newsletter, 'What's New,' on the scientific estate <whatsnew@aps.org>. His 17 June 1994 critique of *Science in American Life* stirred up an Internet debate on the History of Technology, Museum, and Science and Technology Studies discussion lists, among others. Interestingly, virtually none of the participants in the debate had actually seen the exhibition. Within a few months, Park took his campaign to the newspapers, inciting another round of debates in the press. See Park, R L, 'Science fiction, the Smithsonian's disparaging look at technological advancement', *Washington Post* (25 September 1994), and the author's reply, 'Apolitical "science", evidence doesn't support a conclusion of bias in the exhibit', *Washington Post* (16 October 1994). Examples from the science press include Macilwain, C, 'Smithsonian heeds physicists' complaints', *Nature*, 374 (1995), p207; Kleiner, K, 'Fear and loathing at the Smithsonian', *New Scientist*, 146 (1995), p42.

4 At its August 1994 Washington meeting, the American Chemical Society took an informal written poll of members who had toured the exhibit. Of 42 respondents, only one complained of the 'political correctness' of *Science in American Life*, still conceding that the exhibition was 'nevertheless worth doing'. Almost all ACS respondents were highly enthusiastic about the presentation. Poll results are in the author's files, as are supporting letters from other scientists.

5 Secretary I M Heyman publicly stated his views on *Science in American Life* at a National Press Club speech, 23 February 1995: 'I think we should make changes in the exhibition. My guess is we shall . . . I think what happened there, as I might say in my view happened with regard to the original rendition of the *Enola Gay* script, is that there was very little attention paid to what stakeholders would feel—in this case scientists . . ..' See also Stone, A, 'Smithsonian to add "context" to exhibits', *USA Today* (24 February 1995).

6 Dupree, A H, *Science in the Federal Government: A History of Policies and Activities to 1940* (Cambridge, MA: Belknap Press, 1957); Kevles, D J, *The Physicists: The History of a Scientific Community in Modern America* (New York: Knopf, 1978)

7 Doering, Z, Pekarik, A and Bickford, A, 'An assessment of the "Science in American Life" exhibition at the National Museum of American History' (1995), available from the Smithsonian's Institutional Studies Office. The three authors rebut assertions of the exhibition's negative effect in 'Science doesn't scare museum-goers', *Academe*, 82 (letters, May/June 1996), pp3–4.

8 Ringle, K, 'When more is less', *Washington Post* (6 May 1990); Burchard, H, 'Information overloaded', *Washington Post* (11 May 1990); Gibson, E, 'Techno love: Smithsonian sets computer on pedestal', *Washington Times* (10 May 1990)

9 Gross, P R, and Levitt, N, *Higher Superstition: The Academic Left and its Quarrels with Science* (Baltimore: Johns Hopkins University Press, 1994)

10 Sokal, A, 'Transgressing the boundaries: toward a transformative hermeneutics of quantum gravity', *Social Text*, 14 (1996),

pp217–52; Sokal, A, 'A physicist experiments with cultural studies', *Lingua Franca* (May/June 1996), pp62–64

11 An account of this event appears in Gieryn, T, 'Policing STS: a boundary-work souvenir from the Smithsonian exhibition on "Science in American Life' ", *Science, Technology, and Human Values*, 21 (1996), pp100–15.

12 Park, R L, 'Science fiction'

13 See Singer, M, 'Assault on science', *Washington Post* (26 June 1996).

14 Greenburg, D S, 'Thumbs up for science', *Washington Post* (8 July 1996), comments on the paradoxical mood of scientists in the face of good news like the National Science Foundation poll.

15 See McGraw, M-T, 'Curating history in museums: some thoughts provoked by "the West as America" ', *The Public Historian*, 14 (1992), pp51–53.

16 Molella, A, this volume, pp91–97

Collaboration for development and change— or for the sake of funds?

AnnMarie Israelsson

A look at the opportunities for a science centre to collaborate with local organisations.

In 1994 Goéry Delacôte, director of the Exploratorium in San Francisco, and a much respected and long-standing member of the science-centre community, claimed that science centres are an industry on the decline, in obvious need of change and development in order to be able to survive as promoters of scientific literacy.[1] At the time, this was a rather bold statement. Since then, discussion has become more frequent about what could be called 'science-centre ideology', also applicable in parts to science museums, which represents the raison d'être, ideas, and goals in promoting science and technology.

This discussion is more far-reaching than the ideology guiding the educational work we want to pursue, the background of our efforts to reach out to audiences of different kinds, or our more-or-less internal worries over design and content of presentations. We must finally come to grips with the outside world and learn how to relate to it, take advantage of it, collaborate with it, and otherwise involve it in our work to widen and change the scope of our science centres and science museums generally.

Asking necessary questions

Collaboration is, of course, a way of acquiring the necessary financial means for developing and changing science centres and museums. So, take your partners and the dancing will begin! But before we look around for rich dancing partners, I suggest that we try to find out why we should dance, think about what it might lead to, and learn the steps so as not to stumble and fall.

If we want to collaborate to achieve development and change in science-centre activities, we must make sure that we are fully aware of all the aspects of such collaboration, and we must learn to communicate these to our partners: why collaborate?, with whom?, and on what terms? Are there examples of what we should and should not do? Are there shortcuts and pitfalls? What can be gained and what could be lost? What can we learn from each other? I do not claim to have answers to all these questions, but I intend to draw some conclusions from my own experience that may be of interest to others.

Teknikens Hus—an example

It is now 17 years since I started work on setting up a science centre in the city of Luleå in the northernmost region of Sweden. The first steps towards this goal were taken within the local university, where I was at that time the head of the information office. Nine years later, in 1988, Teknikens Hus (House of Technology), opened to the public. The building, measuring 3500 m^2 in size, houses exhibits and activities aimed at promoting public interest in technology. The project was made possible by donations from the big industrial concerns of the region.

Teknikens Hus takes a regional approach to its mission of displaying technology to its visitors—which implies an industrial approach as well as a focus on everyday life. It emphasises the aspects of technology which are close to local life: mining, steel processing, mechanical industry, forestry, hydroelectric energy, communications and technical appliances in the kitchen, the bathroom and the garage. Most exhibits are real objects at work—a lumber machine, a rock drill, an industrial robot (figure 1), a Volvo car, a dishwasher—for which the technology is made visible and accessible so visitors can experiment. Others, for obvious reasons, work on a model scale—a computerised modern saw-mill, a paper mill, a hydroelectric power station—but all are run and controlled by the visitor. There is a sectioned modern house with walls that open to display building methods and materials, water supplies and sewage systems, and electricity installations. The thematic organisation of the exhibits forms a recognisable landscape in which visitors can explore technology in a familiar setting.

Since its opening, Teknikens Hus has been a great success, attracting visitors locally, from other parts of Sweden, and from abroad. More than 170,000 visitors come every year, almost 600 a day—an amazing number for a city of 70,000 inhabitants in a vast and sparsely populated region covering a quarter of Sweden but housing only 3 per cent (260,000) of its population.

Success stories like that of Teknikens Hus are not usually very illuminating. We can learn more from failures than from successes. But let me nevertheless reflect on the existence of Teknikens Hus, and the collaborative work required to start and run it.

Looking at reality

It has always seemed to me that local reality, the local context, has been and often still is ignored in science centres, whereas displaying universal science and scientific principles, first laid out in the Exploratorium at San Francisco, has become a powerful model. Scientists like to teach these universal theories and principles, maybe in the hope that science will become an important tool for everybody to solve the problems they encounter. And yet, a problem posed by reality cannot be solved without taking that reality into account. A theoretical principle cannot be applied without using the cultural alphabet together with the

Figure 1. Although children visiting Teknikens Hus enjoy operating the industrial robot, students from the neighbouring engineering school also use it for training. This was a major incentive for the company ABB to supply the robot and its programming unit.

social and economic structure of the society—the context in which the problem arises or the application is needed. Human problem-solvers need their own experience to be able to put science and technology to good and efficient use. In science centres, however, this experience is seldom appreciated and the proper context is often missing.

Our goal when planning Teknikens Hus was to make technology everyone's concern. We could see no other way to reach that goal than by starting where people are, in their own environment.

Finding a good match

In pursuit of our goal, it would have been obvious to look for support from local technology companies. At first, however, we did not consider taking advantage of knowledge to be found in the environment around us—and we never considered looking for funding there. We asked the physics teachers from the local university for help, but not the practitioners of local industry. Our fund-raising efforts were a Swedish stereotype: we applied for government grants. Sadly, the government answer was a consistent 'no'. But we soon found out that there were other sources of knowledge, and also other funding sources. One reason for our reorientation was that our idea of starting from the practical end of technology did not go down too well with the theoretically minded academics at the university. We found it much easier to talk to people in industry, to ask for professional support, knowledge and skills there. We needed them, and they were pleased to help.

Identifying mutual benefits

Some decision makers in industry displayed a keen interest in promoting technology among young people. In the early 1980s, schools of engineering at institutions such as Luleå University were having difficulty attracting students. The big industrial companies of the region, including a mining company, a steel company and a hydroelectric power company, were worried about recruiting skilled workers for their future operations. It was obvious that elementary schools needed support to be able to teach technology.

Thus, for industry, supporting Teknikens Hus meant supporting a new approach to popularising technology, taking a new educational role, making industry look more appealing to young people and promoting technology on a long-term basis. Here was a way to communicate with audiences that were hard to reach by the traditional kinds of promotional efforts, e.g. industrial fairs or advertising campaigns. Also, it meant an opportunity for publicity with important groups: customers, competitors, employees, teachers and parents. The notion of climbing on to the bandwagon of something new was a strong motivation for some of our supporters.

We learned from this process that identifying the mutual benefits is the first and most important step towards successful collaboration.

Identifying people

The personal attitudes of individual top decision makers have also proved to be of the utmost importance. We were able to persuade one of the most important industrialists at that time to make a large financial donation. Through his involvement, the project acquired a local and regional prestige that proved very effective in convincing other representatives of industry—and politicians—that supporting Teknikens Hus was tantamount to supporting future industrial development.

The forestry industry, which is very important to the regional economy, did not join the other big companies in supporting Teknikens Hus at this stage. Two years later, however, representatives of the forestry industry in the region asked us to collaborate with them to create a forestry exhibition. This was an opportunity we had to take very seriously.

Controversial issues

Forestry is a controversial issue in our part of the world, and was especially so in 1990. The forestry industry was keen to use Teknikens Hus as a means of appearing to be active and environmentally concerned. After all, hydroelectricity—which is and was just as controversial—was displayed in Teknikens Hus and seemed to be defused as an environmental issue in this context.

Most science centres and museums collaborate with a variety of external funding sources to get support for new or special exhibits. This kind of collaboration is well established, although at times it seems to me that science centres and museums are too hesitant about asserting their own identities. Over the years, exhibits treating controversial issues have been created in science centres and museums all over the world—on pollution, nuclear power, industrial waste, etc.—and money has dictated the contents and influenced the design. Exhibitions for science centres or museums have sometimes been produced by industries as a public-relations activity and often in the style of industrial promotion, with glossy pictures, tables, graphs and well-polished pieces of machinery. I would like to see proposals from certain partners being turned down or at least radically changed, to retain credibility.

It may seem a good solution to allow those who have money to work hard and pay to get an exhibition under way. But even if a company tries very hard to detach itself from the subject, its mere involvement will arouse suspicions among the critical public. What role does the science centre really play? Will it do anything for money? The credibility of the science centre or museum is at stake. Once credibility has been damaged, it is very hard to put things right again.

As Teknikens Hus wanted to retain its credibility, the conditions of our collaboration with the forestry industry had to be thoroughly discussed and regulated. We decided to form a joint planning committee with the industrialists to sort out responsibilities and discuss the whole exhibition concept. This proved to be a very effective strategy. Not only did we have access to professional knowledge, but we were able to negotiate how to best use that knowledge, and to confront the views of the professionals. Teknikens Hus took a very clear stand, never accepting anything that disagreed with our aims, even at the risk of losing the financial support.

There were of course disadvantages to this mode of working. The planning committee was large, and meetings were protracted. Negotiations in some cases resulted in omitting certain themes of the exhibition, as we

could not reach agreement. But on the whole, our collaboration worked well. In the course of the planning process a mutual respect developed on both sides. This respect was important to the outcome of the work.

The forestry industry invested SEK4.5 million (about £400,000) in Teknikens Hus—SEK3 million for the exhibit and SEK1.5 million as a contribution to the capital fund for future maintenance. Conditions included an interesting arrangement for ownership of the exhibition: a joint committee of forestry enterprises. This investment was written off by the respective companies over a five-year period, after which we were offered the exhibition at a token price. These conditions were the most favourable to our sponsors with regard to taxation and proved a strong incentive to the committee members to continue keeping Teknikens Hus in mind.

Strategies must vary

Locally and regionally based industrial partners do not necessarily take a common view on collaboration with national companies. After our first funding agreements with the big regional industries, we tried to use the same strategy at a national level: playing either on their sense of responsibility, or on competition.

For instance, although Volvo does not have a production unit in our region, we managed to secure valuable exhibit material and practical manufacturing support from them. A car for daily use was also put at our disposal. SAAB, the other national car-manufacturing company, has a factory making lorry parts across the road from Teknikens Hus. We presented ideas for an exhibition to SAAB, assuming that they would not want their competitors to have an exclusive presence in Teknikens Hus. The answer was curt: the Volvo involvement was no reason for them to become involved. Exclusivity would have been an attractive argument for SAAB to become involved. We decided to start again in the area of aviation, where SAAB has no national industrial competitors. Now we have a very good working partnership with SAAB Aviation, a company based in the south of Sweden, but still no involvement with SAAB car manufacturers.

Small partners and great help

Attractions such as dinosaurs and wide-screen theatres are bound to be of commercial interest, but what about the daily work, the daily needs, the annual running costs? Science centres and museums have different solutions for this, for example charging entrance fees, running gift shops, corporate hire and other activities. This is all very well, except that it may consume more energy than can be justified by the revenues.

Collaborating with small local enterprises offers alternative solutions, creating natural connections between the science centre and the local community and helping to minimise costs. A lot of products needed for maintenance can actually be donated. The pride of a local company in participating in our operation might even be regarded as sufficient reward.

We also allow local and regional sponsors to use the name of Teknikens Hus in their marketing, under certain conditions. For example, one of the local printers produces almost all our information and promotional material. They also print our stationery at no charge. The agreement specifies that they can use a proportion of our printed material to acquire new customers. Being associated with a successful operation like ours benefits their sales.

This way of trying to cover costs is potentially very time consuming. To save time, and also to establish new contacts, we have authorised the entire staff to negotiate and make agreements of this kind in their respective fields, within certain financial limits and provided other staff members are consulted if there is any doubt over the advisability of a deal. As a result of these efforts, our running costs are lower than those of most other comparable operations.

New situations—new partners

Identifying partners for collaboration is probably easier for a science centre specialising in technology like Teknikens Hus than for museums generally. But it is not only exhibition topics and content that can inspire and foster collaboration. Societal circumstances and current developments, both local and national, may also offer possibilities. The prevailing economic recession and unemployment situation in Sweden has presented new opportunities. In 1996 we received a proposal from the job centre of a neighbouring town. Could we receive a group of unemployed women at the science centre as a part of a retraining activity?

Here, we could identify a new group of partners. A special programme is now prepared for groups of unemployed women, starting at their level of knowledge and building on their experience to overcome anxieties about technology that otherwise would impede their retraining for other fields of employment (figure 2). The programme includes practical

Figure 2. Explanations of familiar appliances, such as this dishwasher, are an important part of practical training in technology offered as part of our programme for teachers and for unemployed women

143

technology training for self-confidence, looking at everyday practical problems from a technical point of view and trying out solutions. Our work for the programme is paid for by labour-market authorities. This partnership is very new, and has yet to be evaluated.

I am confident, however, that in this way we can not only form new and favourable partnerships for collaboration, but also take on a wider role in society generally—a role in adult education and training, a role of responsibility in socially difficult situations, and a special role in working towards equality between men and women. In return we shall gain from the input of new audiences with new ideas which will revitalise our activities and change our scope of action.

Collaboration thus is not restricted to access to funding or access to experts. It must also be seen in a wider context as a means of opening up new possibilities and offering new relationships with education and society: a new challenge to the science-centre community.

Notes and references

1   Delacôte, G, 'Science centres: an industry on the decline', presentation to the *Education for Scientific Literacy* conference (London: Science Museum, 6–9 November 1994)

Collections in the modern world

Curatorial challenges:
contexts, controversies and things

Alan Morton

Combining interactives and well-chosen objects from museum collections is a powerful means of exhibiting contemporary science and technology.

A spark chamber has recently been installed in the *Nuclear Physics and Nuclear Power* gallery at the Science Museum, London. This wonderful demonstration was built by scientists and engineers from the Rutherford Appleton Laboratory. But it has a historical relevance, as well as a contemporary one, because the same people built larger versions for particle physics research in the 1970s. Although these chambers are no longer used for research, we have the opportunity of displaying the spark chamber as a demonstration of technical skills that will soon be lost—a high-tech equivalent of a re-creation of handloom weaving in a museum of textile industry. So in answer to the question 'How can we use objects from museum collections to improve the presentation of contemporary science and technology?', I will argue that an exhibition using historic and contemporary objects allows a greater diversity of approach than one which relies primarily on working exhibits, television monitors and text to convey its information.

A strategy that restricts the choice of methods of display, for example by using historic objects or working demonstrations to the exclusion of other devices, obviously restricts the number of possible stories that can be told. That may in fact be a desired outcome, but curators should still be clear about why they choose one option over another. For example, take two galleries at the Science Museum: *Science in the 18th Century* uses historical specimens almost exclusively, and *Launch Pad* is a gallery where interactive exhibits are used exclusively. In *Science in the 18th Century* there is a rolling double cone, displayed as a historical object with no opportunity for hands-on interaction.[1] By contrast,

in *Launch Pad*, the same device, called 'Roll uphill', is displayed without any reference to its history. Perhaps the ideal would have been for *Science in the 18th Century* to have included both an historical example together with a modern replica. Thus the visitor could have the opportunity to try the device, so discovering more about how it works while learning something about the history of science.

In my experience, exhibitions on modern science and technology are more likely to be successful and to engage fully a range of visitors when they include contemporary and historic objects, but also use working demonstrations, reconstructions, television monitors showing archive film, cartoons, photographs and so on.[2] There is no hard and fast distinction between these various categories. For example, historical objects and working demonstrations overlap to great effect in the spark chamber mentioned above.

When we make choices about which objects and modes of display to use in an exhibition, we are not just making decisions about content, but we are also making choices about the approach we adopt to the subject matter and the audience we expect to attract. Consider this case as an illustration of how one individual modified his approach to subject matter and audience. In the 1930s, a physicist in the United States was prevented from working on the Manhattan Project to build the first atomic weapons because he was a member of the Communist Party. Later, during the dark days of the McCarthy era, he was unable to find a job in physics, except as a high-school teacher. One would expect this person to have very individual views about both physics and politics because of his personal circumstances.

This individual was Frank Oppenheimer, whose work led to the founding of the Exploratorium, the hands-on science centre in San Francisco—an apparently uncontroversial approach to teaching science which did not make a feature of historical objects or political comment. The Exploratorium, which became a model for many science centres, was achieved possibly as a result of considerable personal denial of history and politics. It is ironic, then, given Oppenheimer's own background, that this view of science as apolitical and ahistorical was taken up by other science centres and proved to be very popular.[3]

Of course, what was going on was the construction of a very particular view of science, as Simon Schaffer has reminded us.[4] In such cases the absence of historical objects can help to create the illusion that science can exist outside a historical context.

Controversial science

Discussions about how we might display contemporary science and technology are taking place at a time when our subject is changing with great rapidity. Major issues affecting the position of science, and indeed the position of scientists, are being thrashed out. In the late twentieth century, what should we say about science? What image of science should we be constructing/deconstructing at the present time? There are no easy answers. But museum professionals can expect to be caught up in the cross-fire of these battles from time to time. The controversies over the *Science in American Life* exhibition at the National Museum of American History[5] and the *Enola Gay* exhibition at National Air and Space Museum,[6] both part of the Smithsonian, give us much to ponder. In these cases, groups of people holding strong views about the subjects being exhibited took great exception to what was proposed and sought—with varying degrees of success—to influence what was presented.

The so-called 'science wars' in the US provide another uncomfortable example.[7]

Although we have not seen the same passion in the discussions in Europe, we do confront related issues about the value of scientific knowledge, e.g. nuclear waste reprocessing plants, AIDS, bovine spongiform encephalopathy, the building of new highways, science curricula for schools and universities, radioactive waste repositories. In short, we face great challenges in our work when we try to deal with such issues for our audiences. But in museums and science centres we have one great advantage: by providing a commentary on the development of science and technology, we can offer insights to both scientists and non-scientists alike. So how may we use objects to spread enlightenment about science and technology?

I share the concerns expressed by David Lowenthal[8] and others about the difficulties of exhibiting contemporary science when the objects are often visually uninteresting and the subject matter difficult and sometimes threatening. But these difficulties make the challenge all the greater. I am reassured in that much of science and technology, indeed much of everyday life, is shaped by and shapes objects. If objects shape our daily lives, then surely we can use objects to illuminate contemporary science and technology? But it is precisely because many of the objects of contemporary science may be boring to look at, that a variety of approaches is required. First, objects do have an atmospheric, emotive and/or affective value. The point of a museum is precisely to allow someone to visit the past and return to the present. One way to trigger memories and reflection is to display objects which have a personal significance for the viewer: the 'We had one of those!' syndrome. Domestic objects, ephemera and old photographs all provoke these responses. Familiar objects can be used to link the personal experiences of a visitor to the content of an exhibition. I would argue too that the complexity and richness of the subject is diminished if the historical context is omitted.[9]

Although modern technology is important, it is often visually uninteresting. An example of this type of exhibit is two pieces of paraffin wax in the Museum's collections (figure 1). These are the actual pieces of paraffin wax used by James Chadwick to discover the neutron in 1932. In our exhibition, we wanted to make visitors reflect on the fact that such a banal object could change the course of human history. Thus we juxtaposed the object with an interview with Chadwick in *The Times* soon after the announcement of his discovery. The article reported:

Dr Chadwick described his experiments as the normal and logical conclusion of the investigations of Lord Rutherford 10 years ago. Positive results in the search for 'neutrons' would add considerably to the existing knowledge on the subject of the construction of matter, and as such would be of the greatest interest to science, but, to humanity in general the ultimate success or otherwise of the experiments that were being carried out in this direction would make no difference.[10]

Chadwick's views about the significance of his discovery were very wide of the mark: within the next 15 years there was the discovery of fission and the development of nuclear weapons which depended on the neutron. We also made use of photographs and other ephemera to illustrate and make more human the story we wanted to tell.

Macrocosm/microcosm

The treatments mentioned above are not restricted to use with objects of contemporary science and technology. We can also use these techniques to refer to the structures of contemporary science and technology—everything from experimental results, to laboratories, to factories, to research councils. Because objects and organisations are socially constructed, they are all products of their time. Each one is a microcosm carrying the imprint of the macrocosm. Thus, in an exhibition we can explain these connections but at the same time we can use them as a hook to link the familiar to the unexpected or to the otherwise uninteresting black box.

Figure 1. Paraffin wax, with the discs and metal foils, used by James Chadwick in experiments on the neutron

Consider the now ubiquitous barcode which appears simply as stripes accompanied by a number. Before the barcoding system could be introduced into supermarkets, changes had to be introduced in the food industry, food manufacturers and in supermarket chains. The printing industry changed, because of new and more stringent requirements for printing packaging; new technologies such as the laser were used on an industrial scale; jobs were eliminated at supermarket checkouts—and all these changes took place on an international scale. Thus, in developing the barcode, a network of change was involved.[11] Change that is familiar to the public, such as the introduction of the barcode, can be used to capture the interest of a visitor and so draw attention to what else is involved in the broader picture. But for full understanding of the broader changes, insight into the historical processes that formed these structures is necessary.

All-important context

This brings me to one of the great drawbacks of concentrating only on the contemporary: long-term issues and trends are lost sight of, with the result that such exhibitions tend to concentrate on narrow technical issues, and consequently leave out the social and political dimensions of their topic. Nowadays, if an exhibition on nuclear physics was mounted that did not raise moral or social issues, the public—and the press—would be disappointed. If the exhibition was sponsored by an agency connected with the nuclear industry, the public would suspect that the sponsors had had undue influence over the content.

Consider an example from modern particle physics: the most contemporary of contemporary science, so contemporary it has not been built: the Large Hadron Collider (LHC) planned for CERN, the international laboratory for research on particle physics near Geneva. Obviously, there are all sorts of technical issues involved in designing and building such a huge high-tech machine. But recent articles, for example in the journal *Nature*, are not concerned with the technical details of the proposal, instead the discussion is about the financial contributions of various nations. Thus arguments about budget, which in turn raise issues about international diplomacy, gross national products of member states and exchange rates, will be as germane to our understanding of the machine as a historical artefact as the description of how the machine works. When (and if) the LHC becomes operational, the collider project will have been scarred by these budget battles. Nevertheless, it is unlikely there will be a plaque on the wall saying, for instance, that the cost of German unification led to the cancellation of the

detector here or delay in building the magnet there. The physical existence of the machine will be a monument to the heroic team of machine builders and their persistence in overcoming any problems, technical and political. Clearly, such a machine is a piece of contemporary science and technology we would be delighted to cover in museums, and the high-energy physics community would be delighted if we came to have that opportunity.

There is an interesting counter-example, in which the same sort of objects are shown but in a different context. This is the 'Superconducting Supercollider' (SSC) exhibit in the *Science in American Life* exhibition at the National Museum of American History. The counterpart of the LHC in the US was a project to build the SSC in Texas. The project was eventually cancelled. Parts of the accelerator that were built are displayed alongside ephemera illustrating the contributions of the pressure groups who opposed the project. Not surprisingly, as the exhibit draws attention to the inability of the high-energy physics community in the US to carry through their plans, the exhibit has been heavily criticised by physicists who do not like this public reminder of their failure.[12]

Public spaces for science

So what kinds of exhibitions should we put on at the Science Museum? In Britain, there has been a long history of organisations to promote science, from the British Association for the Advancement of Science, through the British Science Guild, to the Committee on the Public Understanding of Science. Does the Museum have to play a similar tune? Or can we do something more? Can we offer a commentary on science and/in society? It seems to me that our role is not clearly defined, but in some ways I am not surprised by this lack of clarity. The point that objects are products of their times applies even more forcefully to museums—which are artefacts too. In the last part of this paper, I will illustrate how exhibitions are products of their time by reference to the work I and my colleagues Sophie Duncan and Chris Berridge are doing for an exhibition to be held in 1997 to mark the centenary of the discovery of the electron by J J Thomson. I will describe the early career of one of the glass tubes with which Thomson discovered the electron as a museum object, in the years when it was an example of contemporary science. To illustrate this, two science exhibitions are referred to: one held before the electron was discovered and the other after, to illustrate how the concerns of those who mounted the exhibitions, or viewed them, influenced the presentation of science.[13]

In the latter part of the nineteenth century there had been wide-ranging changes in science and science education. Physics had become established as a university discipline, with the setting up of physics laboratories for both research and teaching. Many of these reforms had been instigated by the work of the Devonshire Commission in the 1870s. For many scientists of the time, this growth of institutional support for science was not fast enough. Furthermore, in their view, science did not receive the public recognition it deserved, and, in particular, they perceived a neglect of science and technology by the South Kensington Museum (the forerunner of the Victoria and Albert Museum and the Science Museum). The Special Loan Exhibition of Scientific Apparatus held in 1876 was arranged, in part, to remedy this defect. A number of points about this exhibition provide useful comparisons for what came later. The Exhibition showed that:

- science and technology had a history, with instruments and apparatus having an important part in that history
- contemporary science and technology could be illuminated by historical development
- science was an international endeavour in which Britain had an important role.

The exhibition furnished a prototype public space for presenting an image of science. This public space was where scientists, members of the public and commentators could interact.[14]

However, the image of science changed in the next few years. From 1895 onwards there were three remarkable discoveries in as many years: X-rays, radioactivity and the electron. In the decade following, X-rays and radioactivity caused sensations and were covered by newspapers worldwide. Public awareness of physics increased. In 1901, J J Thomson's tube arrived at the South Kensington Museum. The glass tube became caught up in the process of fashioning a public space for science. The construction of a new building for the Science Museum began in 1913, but the First World War delayed its completion. Before the Museum formally opened in 1928, there was another influential exhibition: the British Empire Exhibition at Wembley in 1924–25, at which Thomson's tube was displayed.[15]

Unsurprisingly, the treatment of science in the 1924 Empire exhibition differed markedly from that of the 1876 exhibition, not least because the legacy left by the war meant that the idea of internationalism in science had changed. No longer was science 'European'; it was now portrayed as being explicitly linked to Empire. A further legacy was that the contributions of chemists and engineers to war production was explicitly featured. Interestingly, physics was not prominent, and the contribution of physicists to the war effort went unrecognised in the exhibition.[16] The science galleries were set up under the auspices of the Royal Society in part of the government pavilion (figure 2), and were heavily biased towards sub-atomic physics and physiology. That part of the exhibition dealing with physics was the responsibility of Ernest Rutherford, although much of the organising was carried out by a young physicist, Patrick Blackett. Why it was apposite to link physics with the British Empire is demonstrated by Rutherford's own career. Rutherford had come from New Zealand on one of the first scholarships awarded by the Commissioners of the Great Exhibition of 1851. He had used the scholarship to do research at the Cavendish Laboratory in Cambridge before taking up a post as Professor of Physics in another outpost of Empire, Montreal. Eventually, Rutherford

Figure 2. Part of the Science Galleries at the British Empire Exhibition, 1924, set up under the auspices of the Royal Society

returned to the Cavendish in 1919, taking over from Thomson.

While it seems appropriate for Rutherford to be involved in the Empire exhibition, what about other physicists? For them it was an opportunity to raise the profile of physics. This was necessary because the wartime experience of physicists had been mixed. Though the number of physicists and professional opportunities had grown, physicists felt their status had fallen further behind that of chemists whose standing and career prospects had been greatly advanced by the war. For these reasons, there was discussion about setting up a professional body for physicists and in 1920, the Institute of Physics was set up.[17] Its second president was J J Thomson.

The physics section of the Empire exhibition dealt with the structure of the atom, which was hardly surprising given the interests of its organisers, and began with Thomson's work on the electron. As well as Thomson's tube, the exhibition contained atom models. These are all on exhibition in the Science Museum today, so part of the British Empire Exhibition lives on.

From this brief comparison of the science exhibitions mounted in 1876 and 1924, it can be seen how a public image of science, in particular physics, came to be fashioned. This image was, in part, a consequence of increasing government support for science, but it also reflected the self-image of the new breed of professional physicists about their subject and its uses. Along the way, a new public space was created—the Science Museum. It seems to me that we are now discussing what should take place in these public spaces for science in the future. Are we to display contemporary science and technology simply to augment science teaching in schools? Or, can we be more ambitious and show how science and technology shape and are shaped by modern society? I have argued that by combining objects with modern display technologies, we have the means to do just this. We can develop exhibits that, by providing a historical context for science and technology, are both enriching and informative for a wide variety of visitors, including 'lay people', students and their parents, and also practitioners. The answers to questions about how we may use objects to improve the presentation of contemporary science and technology lie with museum professionals, sponsors and audiences, rather than with the objects themselves.

Notes and references

1 Morton, A Q and Wess, J A, *Public and Private Science: The King George III Collection* (Oxford: Oxford University Press, 1993); Arnold, A, 'Presenting science as product or as process: museums and the making of science', in Pearce, S (ed), *New Research in Museum Studies 6* (London: Athlone Press, 1996), pp57–78

2 For a discussion of how some of these technologies have influenced museum displays, see Morton, A Q, 'Tomorrow's yesterdays: science and technology museums', in Lumley, R (ed), *The Museum Time-Machine: Putting Cultures on Display* (London: Routledge, 1988), pp128–43.

3 Hein, H, *The Exploratorium: The Museum as Laboratory* (Washington, DC: Smithsonian Institution, 1990), pp9–11

4 Schaffer, S, this volume, pp31–39

5 Molella, A, this volume, pp131–37; Molella, A and Stephens, C, 'Science and its stakeholders: the making of "Science in American Life"', in Pearce, S (ed), *New Research in Museum Studies 6*, pp95–106; Friedmann, A J, 'Exhibits and expectations', *Public Understanding of Science*, 4 (1995), pp305–13

6 Wallace, M, 'The battle of the Enola Gay', *Radical Historians Newsletter*, 72 (1995), pp1–32; Harwitt, M, *An Exhibit Denied* (New York: Copernicus, 1996)

7 See Molella, A, this volume, pp131–37, for an explanation of the 'science wars'.

8 Lowenthal, D, *The Past is a Foreign Country* (Cambridge: Cambridge University Press, 1995)

9 An example of the importance of historical context is part of the panel text for the *Nuclear Physics and Nuclear Power* gallery written in light of the controversy over the *Enola Gay* exhibition, US Air Force veterans objecting to what was being said about the dropping of the atomic bombs on Japan and the ending of the Second World War: 'On 14 August Japan surrendered and the Second World War came to an end. For many war veterans and some historians the dropping of two atomic bombs was necessary to avoid an invasion of Japan which could possibly have led to an even greater loss of life, both Japanese and American. However, others argue that neither the use of atomic weapons nor an invasion was necessary because Japan was close to surrender. For these historians, another factor in the decision by the President of the United States to use atomic weapons was a desire to curb Russian influence in the Far East. For on 9 August, the date of the bombing of Nagasaki, Russia was due to enter the war against Japan. Over 50 years later these decisions still arouse painful controversy.'

10 Interview with James Chadwick reported in *The Times*, 29 February 1932.

11 Morton, A Q, 'Packaging history: aspects of the development of the Uniform Product Code (UPC) in the United States, 1970–75', *History and Technology*, 11 (1994), pp101–11

12 Molella, A and Stephens, C

13 For some background to international exhibitions see Greenhalgh, P, *Ephemeral Vistas: the Expositions Universelles, Great Exhibitions, and the World's Fairs 1851–1939* (Manchester: Manchester University Press, 1988); Brain, R, *Going to the Fair: Readings in the Culture of 19th Century Exhibitions* (Cambridge: Whipple Museum of the History of Science, 1993).

14 Habermas, J, *The Structural Transformation of the Public Sphere*, translated by Burger, T and Lawrence, F (London: Polity Press, 1989)

15 Knight, D R and Sabey, A D, *The Lion Roars at Wembley* (London: D R Knight, 1984); *Handbook of the Exhibition of Pure Science Arranged by the Royal Society, British Empire Exhibition, 1924* (London?: 1924?)

16 During the First World War, Rutherford and other physicists from the Empire had worked for the Admiralty's Board of Invention and Research.

17 See Ebison, M G, *The Development of the Physics Profession Linked to the Education and Training of Physicists to 1939*, (University of Salford, PhD thesis, 1991), p134 *et seq.*

Technology at the cutting edge

Edward Wagner

At the Franklin Institute, new technology has been successfully presented to the public in their Cutting Edge Gallery.

Modern technology is evolving so rapidly that most science museums cannot even attempt to present the latest goods that their visitors know are readily available in the shops. With so many products coming on the market every week, it is extremely difficult to decide which ones visitors should see, and an even more demanding task to arrange to display them effectively.

At the Franklin Institute Science Museum in Philadelphia, for the past six years or so, we have been developing ways of presenting new technology to our visitors. The focus of this initiative has proved to be the Albert M Greenfield Cutting Edge Gallery, a demonstration area that features new technological innovations in an informal presentation space, allowing audience members to try devices that have just come on the market.

The Cutting Edge Gallery opened in 1990, in the Museum's future-oriented exhibition space, the Mandell Futures Center. However, while the Cutting Edge Gallery proved successful, the Institute found it difficult to maintain the surrounding exhibition space: we built the future and it didn't work. While the Gallery was designed for a quick succession of live new technology demonstrations, we found it difficult to maintain freshness in the presentation of the static exhibits, which soon became hopelessly outdated.

The Gallery is the only component of the Futures Center that remains. The former Futures Center is now an information technology exhibition called *CyberCity*. Although the exhibition space has changed dramatically, the Gallery continues to fulfil its mission of presenting the latest technologies and making them accessible to our daily visitors.

Through working on these projects at the Franklin Institute, we have learned a good deal over the past few years about the pitfalls and the opportunities of taking new technology to the public in a museum. In this paper, I want to discuss the operation of the Cutting Edge Gallery, including the research, development, presentation and preservation of contemporary technologies. As I shall describe, keeping up with new technology is a considerable challenge to museum professionals, but much appreciated by the visitors.

Why new technologies should be featured

Since its opening in 1934, the Franklin Institute Science Museum has used its live demonstrations to illustrate scientific concepts, including electricity, avionics and chemistry. The Cutting Edge Gallery also gives the museum an opportunity to demonstrate technology. This approach was developed in response to museum studies that showed the importance of presenting applied science in addition to the traditional pure science.

In a report in April 1977 by Herbert Thier and Marcia C Linn, it was observed that 'participatory devices' are generally thought to be more effective than static exhibits in attracting and holding visitors' attention and communicating content.[1] Another author, Minda Borun, concludes that 'successful participatory learning devices allow visitors to manipulate objects in their environment, to conduct experiments, to explore the effects of variation and observe the results'.[2]

In a report published in March 1992, Borun states 'Interaction (19 per cent) and educational value (16 per cent) are the two factors most often mentioned by visitors as contributing to the success/failure of an exhibit. Also important is a sense of personal connection.'[3]

These studies clearly demonstrate the need to have interactive exhibits in a museum. While stand-alone kiosks can offer some interaction, staffed demonstration areas allow the visitor to see the technologies close up. Participatory demonstration areas, such as the Cutting Edge Gallery, provide an opportunity for hands-on experience.

In 1995, the Franklin Institute decided to create an exhibition on computer and digital technology (*CyberCity*). To determine the exhibition's focus, a market-research firm carried out extensive polling of current visitors and potential visitors, concentrating on families. Most poll respondents expressed a strong preference for hands-on experience with computers. The one area that proved to be an exception was the Cutting Edge Gallery. Although the theatre was a hands-on demonstration area, the respondents polled were more concerned with viewing the latest technologies than actually trying them. The Cutting Edge Gallery may be the exception to the rule that respondents would rather participate in hands-on, interactive activities than watch something. The poll results suggested that, after the family had engaged in other participatory activities, watching a demonstration would be a welcome rest and change of pace.

More importantly, the words 'latest' and 'cutting edge' clearly were sources of strong appeal in the printed description of the Gallery. Many people are fascinated by new computer technology. Moreover, respondents were keenly aware of how rapidly computer technology is changing and want to learn about upcoming computer products and how they may be used. One adult visitor suggested that the Gallery might provide a 'wish list' of products his family might like to acquire. However, some concern was expressed over whether the Franklin Institute would be able to update the Gallery frequently enough.[4]

While the Gallery was well received by current visitors, even those individuals who had never visited the Institute or the Gallery before were interested in its approach to technology. These very positive results assured the Institute that the Gallery should remain an integral part of the new technology wing of the museum.

Research and development of new technologies

The term 'cutting edge' usually connotes the latest development in a particular category. The original research for the new Gallery focused on recent developments in research laboratories; however, obtaining and effectively demonstrating these technologies became extremely difficult. To solve this problem I developed the following definitions:

Absolute cutting edge. Technologies on the leading edge of research. These include those still in development, being beta-tested or just released. Examples include high-definition television (HDTV) and hydrogen-powered planes.

Relative cutting edge. Technologies that have been developed within the past three to five years but are not readily accessible or well known to the general public. Examples include assistive technologies for people with disabilities, such as a breath-controlled wheelchair and voice-synthesizer systems (figure 1).

While featuring equipment on the 'absolute cutting edge' of technology is very exciting, visitors react more favourably to 'relative cutting edge' presentations. Visitors report that they can better identify with technologies that already (or will shortly) affect their lives. Therefore, the Gallery's presentation schedule includes a balance of both categories, with an emphasis on the 'relative cutting edge' technologies.

Figure 1. Demonstrating a voice-synthesizer keypad

Acquisition

Since the Gallery's annual budget is extremely limited, I have enlisted the support of corporations in obtaining exhibition materials. To market the Gallery to the companies, I pointed out the advantages of having a product on display. Instead of operating under the opinion that the participating companies were doing the Museum a favour by lending us the equipment, I put it to the companies that we were doing them a favour. In return for lending the Museum their equipment, the participating companies benefit from months of free advertising. The Gallery is able to display the latest in technology at negligible cost, and to continue to update the features as new equipment is developed and released. Participating companies usually pay for shipping, installing and maintaining equipment in the Gallery. If necessary, they provide technical support. Companies also sometimes supply guest presenters for special 'event weekends'. This 'package' is very successful, and companies now often contact me to suggest new products that could be presented in the Gallery.

Prioritising displays

Since the technologies considered for demonstration have different potential visitor and public relations appeal, their potential features are classified in the following way.

Major features. Technologies/products that are both dynamic and of great interest to all visitors are classified as major features. The devices presented are often not yet available to consumers (such as HDTV or hydrogen-powered cars). These presentations are usually on display for two to three months and attract large attendances. They also provide excellent opportunities to showcase the Gallery for public-relations purposes. Projects of this scale are presented two to four times a year and given priority scheduling dates. An example of this type of demonstration is 'Cutting Edge Vision', which featured technologies used by individuals with little or no vision to navigate through their environments. The demonstration also included enhanced vision technologies, such as infrared sensors and night-vision goggles.

Minor features. Technologies/products that, while interesting and dynamic, are not as impressive as major features are classified as minor features. These presentations may be technological breakthroughs that are now available to the visitor but are still relatively unfamiliar. The features are on display for one to two months and are scheduled around the major features. An example of this type of demonstration is the compact-disc interactive presentation, which featured compact-disc technology developed by Philips. This demonstration focused on one of the many interactive disc-based systems available commercially.

Auxiliary exhibits. Technologies/products owned by the Gallery are auxiliary. These are used to provide continuity of exhibits in the

157

Gallery in case of unexpected delays, for example, in the set up of a feature. These presentations can be quickly set up and demonstrated when necessary. An example is a demonstration of the videophone, which featured an available technology that most visitors had not seen before and that continues to be of interest to them.

Criteria

Because the Gallery does not focus on a particular technology area, the number of possible features is nearly limitless. Through trial and error, certain criteria have been developed to help focus the research.

The importance of demonstrating actual equipment (not mock-ups, simulations or only images) became apparent after attempts to demonstrate technologies such as magnetic levitation (maglev) trains and nanotechnology (such as microcircuitry) were unsuccessful. While audiences showed interest in the subject areas, there was no way to present the information in an engaging manner. In the case of maglev trains, mock-ups and video footage were used because the actual trains would not fit into the Gallery. For nanotechnology, while the real devices could be shown, the actual operation of the chips could not be seen by the audience. The audience was required to rely on the demonstrator's verbal explanation and (again) video footage. Examples of successful demonstrations of actual devices include HDTV, virtual reality and solar panels.

Devices that visitors can try themselves are also important. Although the *CyberCity* poll showed that visitors are content to watch a demonstration of new technologies, observations by staff showed that some audience participation is important to the success of a demonstration— especially if there are children in the audience. The staff feel that this involvement increases an audience's interest in the demonstration and produces greater enjoyment. Examples of successful demonstrations in this category include digital movie editing, computer-based temperature probes and videophones.

Through observation and informal interviews it was concluded that features on emerging technologies that had no apparent benefit to the visitors were not well received. Certain adult visitors (especially those with no science background) were fascinated by these new technologies but most visitors felt that the technologies had no relevance to their lives. On the other hand, visitors reacted more favourably to features that they felt related to their own lives or individuals that they knew. This was true of features demonstrating:

* technologies entering the consumer market, e.g. personal digital assistants, or the commercial market, e.g. virtual reality.
* technologies with apparent consumer or commercial applications, e.g. the global positioning system, which can display its location anywhere in the world based on information from satellites.
* technologies with important special interest audiences, e.g. electronic innovations for the physically challenged. The Gallery has featured demonstrations on different aspects of assistive and adaptive technologies annually during the past four years. These features have proven extremely popular with both visitors and staff.

Topic selection

The number of possible new features is large, but operating constraints (theatre size, budget, and safety) and general audience appeal aid the Gallery staff in focusing their research. As there are typically five different presentations a year, the staff focus on technologies that they feel will have the most audience interest. In addition, certain technologies use equipment too complex for visitors—or staff—to understand fully during a brief demonstration. Because the demonstration staff, who have a variety of science backgrounds, are assigned to demonstrate the technology and answer questions about it, it is important for them to have a clear understanding of both the device's operation and the demonstrated principle. Many visitors expect the demonstrators to be experts in whatever field they are demonstrating, so it is important that they become proficient in the featured technology within a reasonable training time. Examples of technologies considered too complex for demonstrations include certain areas of biotechnology, nanotechnology and low-level computer operations (e.g. circuitry).

Some technologies have safety, power or handling restrictions that require excessive training periods and frequent procedural reviews of demonstrations. They may also involve long pre- and post-demonstration preparation time. While the time involved would be acceptable if one or two individuals were dedicated to this type of demonstration, it is not practical for the general demonstration staff. In addition, some demonstrations may require major modifications or additions to the Gallery that are not within the Gallery's budget. Examples include technologies using radioactive material, those requiring massive power consumption and those involving expensive and/or fragile equipment.

Timeline

Topics for new presentations are found using a variety of sources: technical literature, trade journals, conventions and exhibitions, by developing and maintaining professional contacts, and by browsing the Internet and World Wide Web. Features are selected six months to a year before they are scheduled to be presented. This allows sufficient time to contact vendors, arrange in-house public relations and coordinate demonstration development.

When the Gallery opened in 1990, it featured a new demonstration each month. While this frequent turnover maintained the feel of a dynamic theatre space, most individuals do not visit the museum monthly and may miss many interesting features between visits. In addition, the short run required the Gallery staff to be simultaneously researching,

developing, training, maintaining and demonstrating the features. The demonstration staff usually require two weeks to learn the new demonstration while concurrently presenting the existing demonstration, as well as performing demonstrations in other areas of the museum. The Gallery now presents new features every two to three months. This allows visitors more opportunity to attend each new demonstration and enables the Gallery staff to better manage the development and delivery of the features.

Presentation

Because the audiences include school groups, families and adults, demonstrations need to incorporate different levels of technical information. The duration of a presentation is tailored to the audience: usually 20 minutes for children and 45 minutes for adults and families, which allows for questions and interaction. The seating area accommodates about 50 people and faces a small stage where the demonstrations take place. Because the demonstrations change so frequently, devices may not be in as good condition as those displayed in the surrounding Gallery. However, the audiences appear to take this into account and we have never had negative feedback on this aspect.

Retaining and preserving equipment

The Gallery generally does not keep devices after they have been featured. Larger companies often limit the length of time the equipment is on loan. Smaller companies usually do not have the capital to permit permanent donations. However, the Gallery has had numerous pieces of equipment donated, ranging from inexpensive temperature monitors to very valuable computers. Most of the equipment is placed in storage, because the systems are occasionally displayed inside static display cases during other features, or used to demonstrate other technologies (in the case of computers or monitors).

159

Recently, the Museum's curatorial staff have begun an effort to identify and tag all artefacts within the Museum. This provides the ideal opportunity to move the donated devices not currently in use into the curatorial storage area.

Some of the Gallery's computer-related acquisitions are to be used in a new *CyberCity* exhibition area, the 'Antique Shop'. It was found, however, that those who specified an interest in the history of other technologies such as cars and planes shrug their shoulders at the history of computers.[5] The computer and other emerging technologies, they claimed, focus on the future, not the past.

Another possible explanation for the lack of interest in the history of computer technology is the fact that there are no apparent differences between generations of computers. As the differences are usually found inside the computer's microchips and are not readily visible, it is difficult to understand what is happening and what has changed in the various models. While older technologies, such as trains and printing presses, usually have some visible movement, a view of the microchip provides no hint as to what is going on inside it.

Like computer hardware, computer software is also difficult to comprehend through visual inspection. A CD-ROM of the *Bible* looks the same as one of a computer game. The difference is only apparent when the information is seen on the computer screen. The subtle differences in generations of computer hardware and software also pose problems for curators who have to catalogue the equipment, place it properly in the Museum's collection and possibly curate a future exhibit featuring it.

Conclusion

The Cutting Edge Gallery was conceived of as a demonstration space within the Franklin Institute where visitors could experience emerging technologies in a hands-on environment. In its first six years, the Gallery has featured over 50 different technologies, from oil-eating microbes to virtual reality. Many visitors have commented that their first contact with these emerging technologies was at the Gallery, and that they were pleased that the Institute provided this service.

Studies show that visitors prefer hands-on, interactive exhibits and demonstrations. The Gallery is able to offer this sort of demonstration on a daily basis. The demonstration staff use the developed script to interpret the technology and, if possible, allow the audience members to try the devices. This combination of live interpretation and hands-on experience has proved to be very popular with visitors.

The curatorial staff are now working with the Gallery staff to preserve items donated to the Gallery and to assemble them into comprehensive collections that can be presented in the future. However, as technologies become smaller and computers are integrated into their operation, presenting them in a demonstration becomes more challenging because explaining how a computer operates is difficult. Explaining, too, how a new technology uses a computer to perform other functions is even more difficult. The Gallery presentation staff need to become competent in both computer technology and the specific technology being presented. When the feature is finished, the Gallery staff need to work with the curatorial staff to make the curators aware of the technology's significance. The curators can then correctly position the equipment within the Museum's collections.

The Gallery continues on its mission to interpret these emerging technologies by explaining how the technologies work and their relevance to visitors' lives. Families, groups and individuals visit the Museum daily to see and learn about these new innovations. It is the responsibility of the Gallery's demonstration staff to present this information effectively on technologies at the cutting edge.

Notes and references

1 Thier, H and Linn, M C, 'The value of interactive learning', paper presented at *Education and Science Centers Workshop* (Boston: Association of Science Technology Centers Meeting, 1975), p66

2 Borun, M, *Measuring the Immeasurable: A Pilot Study of Museum Effectiveness* (Philadelphia: Franklin Institute Science Museum and Planetarium, 1977), p67

3 Borun, M and Chambers, M, *Roles of Affect in the Museum Visit and Ways of Assessing Them*, *vol 3* (Chicago: Museum of Science and Industry, 1992), pp7–19

4 DePaulo, P J, *Testing the Concept of Cyber-City: Qualitative Interviews with Visitor and Non-Visitor Families* (Montgomeryville: J Reckner Associates Inc., 1996), p14

5 DePaulo, P J, p15

Paradise and Pandora's box:
why science museums must be both

David Lowenthal

The challenges for museums lie not in what to collect or conserve, but how to retain an audience increasingly remote from science.

Science museums face two conundrums peculiar to science itself. The first is a growing invisibility of scientific research that renders any display of artefacts ever less meaningful. The second is mounting concern that, far from simply improving the human lot, science also creates parlous vexations, some of which imperil our very existence.

First, invisibility. Much science today is either microscopic or megascopic, observable only by specialists with access to powerful and expensive apparatus. And its fruits emerge by technologies remote from any eye. Such past discoveries as electricity, radio waves and magnetic current were obscure; computers and the Internet are utterly opaque. Electronic systems that alter contrast, colour and density, notes Brian Bracegirdle, show us selectively detailed images 'that even *we* cannot see in our microscopes'.[1] Science today becomes ever more mysterious.

Let me illustrate. I have shown how the American scholar–diplomat George Perkins Marsh came in the 1860s to write *Man and Nature; Or, Physical Geography as Modified by Human Action*, the pioneer work of what is now called ecological insight.[2] These insights stemmed from Marsh's studies of environmental change in New England, the Middle East and Alpine Europe. In Marsh's day and for some time afterwards the harmful effects of human agency were patent to the eye—erosion, siltation, unstable and extreme river regimes.

Such impacts remain highly visible. Somewhat less visible are Rachel Carson's demons—pollution, DDT and other chemical poisoning.[3] But such risks, if far from overcome, are not those that now most alarm us. None of the key issues discussed in Rio in 1992—acid rain, stratospheric ozone depletion and global warming—had been of major significance in Stockholm only two decades earlier. Along with nuclear fallout, threats before unheard of now command our severest concern for survival.[4]

These new risks have one feature in common: their virtual invisibility. They are doubly fearsome because they are perceptible only by arcane experts. And even experts dispute how much is seen: the Madrid intergovernmental working group on climatic change in 1995 barely agreed to call human influence on global climate 'discernible' instead of 'appreciable', 'notable', 'measurable' or 'detectable'.[5]

Such impacts typically become visible only when it is too late to reverse their effects. According to what one analyst has termed 'the paradox of insignificant change', a sequence of 'innumerable seemingly meaningless changes ultimately accumulate [and] erupt in cataclysmic proportions'.[6] Equally, we ruin the planet through 'small events in which we all participate: the spraying of lawns, the dumping of oil, the demands for wormless apples, beachfront cottages, layers of packaging'.[7] Vainly seeking evidence of a much-heralded fisheries catastrophe, Michael Parfit found something 'less immediate and more complex . . .. The most frightening thing . . . was not the cod crash off Newfoundland but something a skipper told me in Dakar: "The fish just get a little smaller each year." The unspectacular change takes the biggest toll'; but we seem to respond only to spectacles of imminent doom.

The second conundrum is the Enlightenment dream of perfectibility that saddles science with a self-justifying imprimatur of rational progress. The birth and growth of science are indeed coterminous with gospels of progress and reason. Although science has always had its detractors, it was widely if not universally extolled until sometime between the start and the middle of this century. Today the scientific enterprise is widely feared and resented even by those who take its benefits for granted.

Science is feared and resented both because its mysteries make it remote and authoritarian and because its unintended consequences seem increasingly ominous. The risks of nuclear power are typical. 'Once considered a glittering technological panacea', find environmentalists, nuclear power shows 'how the most radiant innovations sometimes cast the darkest shadows of risk. Public distrust [mounted] as worries accumulated' over economic, health and safety costs. 'Seeds of suspicion, . . . fertilized by civic empowerment and cynicism' have gone far to strangle nuclear power in the United States.[8] It is crucial that scientists acknowledge, and science museums address, these widespread fears and resentments.

The magnitude of the shift in public sentiment is patent in earth science. Persuaded that the forces of nature vastly exceeded those of man, geologists and engineers long expressed faith in progress. Whatever men did, they could not seriously harm nature. However far technology extended environmental impacts, man would remain only a minor geological force; the gravest man-made disasters meant only small and temporary setbacks in his progressive mastery of an infinitely resourceful Earth.[9]

By the early nineteenth century, observers everywhere had detailed mounting environmental damage owing to deforestation, intensive farming and other causes. But these untoward effects were ignored by policy makers. Evidence of impact went largely unheeded because its scattered and abstruse sources were uncollated. Meanwhile, cornucopian progressivists went on assuming that

science could safely enlarge its power over malign raw nature, while environmental determinists left ultimate power still safely sheltered in nature's might.[10]

Thus as late as 1929 Freud could extol the conquest of nature in these terms:

We recognize that a country has attained a high state of civilization when we find . . . everything in it that can be helpful in exploiting the earth for man's benefit and in protecting him against nature. . . . In such a country the course of rivers . . . is regulated . . .. The soil is industriously cultivated; . . . the mineral wealth is brought up assiduously from the depths; . . . wild and dangerous animals have been exterminated.[11]

Many, perhaps most, still view technological impacts as largely benign. But it is a measure of how much less sanguine we have become that, only two generations on, Freud's accolade to technological progress now strikes us as astoundingly crude and human-centred. Such blithe neglect of technology's negative consequences is today scarcely conceivable.

Yet many scientists and even science curators cling to implicit and often explicit faith that the history of science is a story of success—success in unlocking the secrets of nature, in knowing it with ever greater certainty, and in continually augmenting its management for human benefit.

For example, one issue addressed in this volume is whether museums should engage with science so current that it is still under debate among scientists. The query itself can be profoundly progressivist: it implies that science is a march towards confident consensus, the great results of which are best paraded as uncontroversial. In this triumphalist reading, *current* science becomes too debatable to be a proper subject for public display and discourse.

No other museum realm today would begin by assuming that dissentient views were perhaps too problematic to tackle. In history, archaeology and art history, perceived truths are continually revised by the finding of new materials, the insights of ever-unfolding hindsight, and the social politics of differing knowledge communities. Only science keeps faith in fixed and universal truths and views the history of change as one of cumulative advance.

These two conundrums intensify each other. What is invisible must be taken on faith. And scientists rely on public faith in their own credibility and in the untarnished virtue of their enterprise. But such faith is fast eroding. Science more and more comes under assault for the inadequacy or perversity of its consequences. Irrational suspicion of science is rife among even the supposedly well educated. Half the students in an American college astronomy course in 1995 thought UFOs credible, and suspected government of concealing known existence of aliens; such views stemmed wholly from televised entertainment.[12]

The dire effects of seemingly benign advances suffuse scientific enterprise. Pesticides create resistant superpests and antibiotics make tougher microbes; we risk running out of remedies for plagues once supposed forever subdued. Repetitive strain injuries such as carpal tunnel syndrome are now epidemic among keyboard operators. And we replace acute problems with chronic disabilities. High-speed medical evacuations greatly reduced head-injury deaths among American soldiers in Vietnam—but as a result many more with severe brain damage need day-by-day care.

'Each breakthrough has . . . created problems that tend to defeat the reason for adopting the technology in the first place', concludes Edward Tenner. Tenner thinks we can learn 'to recognise unfortunate consequences early enough to counter them'.[13] But increasing invisibility and time-lapse between cause and effect seem to me to presage less auspicious futures.

One example will suffice. The United States Environmental Protection Agency plans to bury nuclear waste beneath a thousand feet of rock at Yucca Mountain in southern Nevada, on the assumption that container safety against toxic residues can be assured for 10,000 years. But this materials-based projection ignores changes in human society and behaviour apt to impinge on the site. Such changes are unknowable one century ahead, let alone ten millennia.[14] Moreover, even were government agencies, notoriously unable to plan ten years ahead, now to be relied on for 10,000, this would not be nearly long enough. For sufficient decay of radioactive isotopes—potentially lethal in air or groundwater were packaging to leak—requires as much as a million years. To such dilemmas, present-day science offers no solution whatever.[15]

Science is bound to become less accessible, more dubious in its outcomes, and more detested by a mystified and beleaguered public. Genetic progress is a case in point: to know the statistical prospect of inheriting some fatal or crippling disability requires agonising decisions by patients and potential spouses, employers and life insurers.[16] As elsewhere in science, we confront Faustian dilemmas—choices between destructive knowledge and wilful ignorance.

These conundrums lie at the heart of science as seen by the public. Science museums must tackle them or lose touch with vital concerns about the impact and continuing role of science. The issue is not what things to collect or conserve—it is how to address audiences who retain fascination with, and even residual faith in, scientific progress, but who know ever less how science works and turn increasingly to anti-scientific faiths out of ignorance and dread of Pandora's box. A display tracing fears of 'Science as Satanism' from the seventeenth century to the present might be one place to begin.

Notes and References

1 Bracegirdle, B, 'What our forefathers did *not see!*', *Science Museum Review* (1987–88), pp28–30; Lowenthal, D, 'Science museum collecting', in *Museum Collecting Policies in Modern Science and Technology* (London: Science Museum, 1991), pp11–15

2 Lowenthal, D, *George Perkins Marsh, Versatile Vermonter* (New York: Columbia University Press, 1958); Marsh, G P, *Man and Nature; Or, Physical Geography as Modified by Human Action* [1864], Lowenthal, D (ed) (Cambridge, MA: Harvard University Press, 1965)

3 Carson, R, *Silent Spring* (London: Penguin, 1962)

4 White, G F, 'Emerging issues in global environmental policy', *Ambio* (February 1996)

5 IPCC proceedings reported in Sawyer, K, 'Panel: humans affect climate', *International Herald Tribune* (2–3 December 1995)

6 Massa, I, 'The paradox of insignificant change perspectives on environmental history', *Environmental History Newsletter*, 5 (1993), pp3–14, ref. p11

7 Parfit, M, 'The threat is incremental harm, not catastrophe', *International Herald Tribune* (19 December 1995)

8 Pasqualetti, M J and Pijawka, K D, '*Un*siting nuclear power plants: decommissioning risks and their land use context', *Professional Geographer*, 48 (1996), pp57–69, ref. p57

9 Gregory, K J and Walling, D E (eds), *Human Activity and Environmental Processes* (New York: Wiley, 1987), p3

10 Glacken, C J, 'Changing ideas of the habitable world', in Thomas, W L Jr (ed), *Man's Role in Changing the Face of the Earth* (Chicago: University of Chicago Press, 1956), pp81–88; Lowenthal, D, 'Awareness of human impacts: changing attitudes and emphases', in Turner, B L II, Clarck, W C, and Kates, R W, *et al.* (eds), *The Earth as Transformed by Human Action* (Cambridge: Cambridge University Press, 1990), pp121–23

11 Freud, S, *Civilization and Its Discontents* [1929], translation by Rivière, J (London: Hogarth Press and the Institute of Psycho-analysis, 1946), pp53–54

12 Anderson, W R, 'A federal UFO cover-up? Yes, they saw it on TV', *International Herald Tribune* (30 August 1996)

13 Tenner, E, *Why Things Bite Back: New Technology and the Revenge Effect* (London: Fourth Estate, 1996). A preview of this book appears in 'Machine bites man', *New Scientist* (27 July 1996), pp48–49.

14 Erikson, K, *A New Species of Trouble: Explorations in Disaster, Trauma, and Community* (New York: W W Norton, 1994), pp203–25

15 Fri, R W, 'Using science soundly: the Yucca Mountain standard', *RFF Review* (Summer 1995), pp15–18: summarising National Research Council, *Technical Bases for Yucca Mountain Standards* (Washington, DC: 1995).

16 Kevles, D J and Hood, L (eds), *The Code of Codes: Scientific and Social Issues in the Human Genome Project* (Cambridge, MA: Harvard University Press, 1992)

The virtual visit

Reinventing museums through the information revolution

Oliver Morton

New developments in the information age will allow museums and science centres to reinvent themselves.

The speed and extent of progress in digital communication make it today's most powerful and pervasive area of technological change. It is easy to treat the immediacy associated with this digital revolution as a definition of the contemporary, of what it means to be now—and such a definition might seem to leave little room for the abiding characteristics of the museum. But this would be a misreading. To be contemporary is not to be obsessed with novelty, to be unendingly up-to-the-minute, to be in constant flux. Understanding this permits interesting insights into the possible relationship between digital media and museums.

The most beguiling thing about the new possibilities represented by, but not limited to, the Internet is the explosion in the ease of communication, of access, of immediacy. The fact that it is possible to send messages, sound files and images easily, cheaply and close to instantaneously around the world, or to make them available for fetching at similar speeds, is immensely attractive. Imagery of speed, of riding the wave, of being at the cutting edge, of having one foot in the accelerated future, permeates the culture of the Internet. In this, that culture reflects and exaggerates a more widespread fear of being behind. In the traditional media, the declining daily newspapers are moving from breaking news to discussing it, analysing it and supplementing it in a way that is traditionally the preserve of the magazine. They are ceding their indispensability to the electronic media which have succeeded in capturing the immediacy of now, in particular the 24-hour rolling news services. Internet media are going the same way, increasingly

priding themselves on 24-hours-a-day, seven-day-a-week updating. 'Streamed' media (continuously changing data such as those in a piece of music or imagery) are becoming technically possible on the Internet: combined with the notion of 'push', whereby data are sent unsolicited by a media provider, the Internet would seem to be becoming ever more immediate, ever more of the moment, an exaggeration of contemporary media forms rather than an alternative to them.

However, immediacy, while attractive, is not the only thing the digital media have to offer. There is also storage. Data-storage capabilities have grown as quickly as data-processing capabilities. The archival possibilities of digital technology have grown as quickly as the transmission possibilities, or quicker. And, at the moment, the single most popular application of the Internet, after e-mail, is an archival one: the World Wide Web, which depends on a Hypertext Markup Language that is a direct descendant of the archivists' Standard Generalised Markup Language. The ability to store and cross-reference material on the Web, which links millions of computers across the world, is close to inexhaustible. And within this archive the pressing urge of the moment common to other media, the relentlessness of 'now', is diminished if not lost. This immersive and widely shared archive allows the possibility of new forms of public information, new forums, perhaps, for public debate, in which museums might be able to find a helpful new role.

Take an example from my profession, journalism. The typical information form is that of the news article. A typical archive is a clippings file. Neither provides fully satisfying ways of informing oneself about a complex issue of interest, such as, to take a fairly contemporary example, the farrago of agriculture, epidemiology, theoretical biology, pathology, international relations and public health that goes by the name of 'the BSE crisis'. Each specific article is forced to present what is new first and foremost, and then fill in relevant bits of background according to that news. It will also tend to reflect only one of the possible takes on the issue: the political take interesting to the opinon pages, the beef-futures take of the financial pages, the prion take of the science pages. To get the whole story to make sense is a tedious exercise in newsprint stratigraphy. It involves discounting huge amounts of repetition and a certain amount of frank contradiction as the two-dimensional projections of 'nows'-past are assembled into something more satisfyingly rounded and complete.

In the not-necessarily-immediate world of on-line data, a different sort of communication becomes possible, something that may not even be journalism at all, but some other form of presentation. Imagine an article that synthesises all the relevant events without privileging the most recent, which is updated regularly but which does not use the latest update as the peg for its appeal. It assumes that you want to know the whole thing in a certain amount of detail: it refers you elsewhere for some niceties and subtleties. It grows over time, it changes over time, but no moment in time is its master.

Using the traditional media, which are all about getting information from producer to consumer on a regular basis that can be charged for (charged to either the consumer or the advertiser), there is no clearly economically viable way to make such an artefact available. On the Internet, where marginal distribution costs, over and above the costs of making something accessible in the first place, are effectively nil, one could leave it and let people discover it in their own time. Its

asynchronous audience would grow and wane over weeks and months, but its running total might well grow to be impressive by anyone's standards, even though on any given day far fewer people might see it than would see that day's 'news' coverage of the same issue.

One of the exciting things about the revolution in information systems that we are going through is that it redefines the boundaries of the media; it allows almost any institution to take on a publishing role. In time, it will force them into something like that role. There will be little future for any body that does not wish to make itself known through the digital media. And new publishing challenges, new media forms, may be better exploited by non-traditional-media players than by the established publishers and broadcasters.

In scientific matters, museums might be ideal homes for the new form of information sketched out above, carrying as they do a sense of the archival, of the authoritative, of the disinterested. The arrangement of information and the assessment of relevance are, or should be, key skills for any curator. The existing collections of a museum (or of many museums, working together) could be articulated into virtual arrangements that add depth to exposition; non-museum resources could be used similarly.

Perhaps most promising, from this point of view, is the fact that museums enjoy relationships with their users unlike those which typify the media. Museums are accessed simultaneously and independently by large numbers of people who use the material on display in different ways—ways that can, but need not, influence the choices made by others also present. Many media are accessed simultaneously; some, such as books, are accessed simultaneously and independently, with each user controlling the interaction; others, like film, are accessed en masse in ways that allow the members of the audience to influence each other, but cannot be related to independently. Museums and exhibitions provide a unique possibility of interaction and independence, of shared experience and unique reactions.

This free-floating pattern of use, if it can be transferred or extended into the digital realm, seems a far more promising basis for debates and discussions than the feedback-to-the-originator 'letters to the editor' model of the traditional media. Some on-line experiments that mix debate, analysis and audience feedback are showing a way forward towards such a goal. The key, I suspect, will be extension—to make the on-line space in which an issue is made available feel like a continuation of a gallery space, but one in which direct communication, rather than observation and emulation, typifies the way users interact.

The digital revolution is not about speed or processing power; it is not about synchronising the world, or streamlining the progress of traditional data forms. It is about information and interaction. As such, it presents an opportunity for the museum to reassert, or create afresh, a relevance it does not now enjoy.

The virtual visit: towards a new concept for the electronic science centre

Roland Jackson

Internet-based technology will enable museums and science centres to make a transformation from top-down to bottom-up communication with the public.

The vision

Where does the public find out about contemporary science and technology? Is it from magazines like *New Scientist* and *Scientific American*? From features in *Le Monde* or the *Independent* newspaper? Or perhaps from radio and television programmes? Almost certainly, science centres and museums have played a relatively small part—so far. The advent of the Internet gives us the potential to make a much larger impact. Unlike magazines, newspapers and television programmes, the basis for our activities lies in the construction of relatively long-lasting exhibitions, supported and complemented by debates, discussions and other means of exchanging ideas and views.

Given their focus on exhibitions, science centres and museums now have a natural opportunity to lead the process of developing electronic exhibitions and the associated means of information exchange. Perhaps no other type of institution has the same clear potential niche or mission. Science centres and museums should be developing as nodal points in the access of the public to contemporary science and technology. They should be mixing thematic electronic exhibitions that have immediate updates with live on-line provision of information. They should also be mediating and guiding public access to the discoveries and issues of the day, and to the people, views and concepts behind them. Whether it is the possibility of life on Mars, the vulcanology of Iceland, the safety of eating beef, or the progress and potential of the Human Genome Project, electronic exhibitions created by science centres and museums ought to be the natural starting points for the public. My proposal in this paper is for a new paradigm to shape our work with the public towards this end.

The problems of prediction

The prediction of the impact, potential and development of electronic technologies is a risky business. Implicit in this paper is the belief that global digital networking systems will change the nature of science centres and museums and their relationships with society, or at the very least with some parts of it. If I am completely wrong, at least I shall be in good company, as the following quotes demonstrate (even if they are apocryphal and probably taken out of context):

'I think there's a world market for maybe five computers.'—Thomas Watson, chairman of IBM, 1943

'There is no reason why anyone would want to have a computer in their home.'—Ken Olsen, president, chairman and founder of Digital Equipment Corporation, 1977

'640K ought to be enough for anyone'—Bill Gates, chairman and chief executive officer of Microsoft Corporation, 1981

I conclude from those comments that one should err on the side of boldness.

Concepts and starting points

The emphasis in this paper is entirely on the Internet and related technologies, rather than stand-alone systems, dedicated networks within science centres and museums, or specific digital products such as CD-ROMs. The important features of networking technologies include that they allow real-time and interactive access by multiple simultaneous users, from (and to) any computer anywhere in the world that has a connection to the Internet. All three parameters are significant.

Real-time and interactive access. The virtual world is organic and can be up to date in a way that, for example, a book or a CD-ROM cannot.

Multiple simultaneous users. Of course it is possible to have multiple-way phone conferences, and multiple accesses to local computer networks, but global networking adds a qualitatively different degree of scope and, in particular, of scale.

Any computer anywhere in the world. This remote access is the real distinguishing feature of the global networks, allowing instant communication across and between different people, cultures and information sources.

Built on these three physical features of the system are the opportunities for the 'visitor' or on-line participant. These too can be grouped into three areas: access to data and information, access to conferencing and discussion, and access to what I shall define as collaborative computing applications.

Before examining each of these in turn it is necessary to consider the ways in which the public at large might make use of global networking technologies as they develop. For our purpose—that of presenting contemporary science and technology—three distinct groups of people are identifiable.

First, there is a not negligible group that will make little or no use of these technologies for leisure or even for general educational purposes. We might classify these people as technophobes, or simply recognise that they have different preferred styles of interacting and learning. Such people find the multimedia screen unidimensional (indeed not multimedia at all), lacking in personal warmth and the reality of the physical world. I suspect this group is, and will remain, a large group, and

will hopefully continue to flood into our physical environments.

Second, there is a group of people who use the Internet as a library resource and as a means of exchanging views, but in the context of information structured largely by others. These are the people we can attract to our on-line exhibitions and conferences. We certainly do not yet know how large or diverse this group might be, neither do we know the details of the ways by which it uses our existing provision. There is a desperate need for more user research, given all the experimentation currently in process.

Third, there is an emerging group of people who want to create new electronic worlds with others, including those who have an interest in shaping presentations of science and technology according to their own views and interests. The new technologies give us ways of empowering people to work creatively with like-minded enthusiasts to these ends, and I believe that science centres and museums should support, and indeed lead, this process. This group of people is, and may remain, small, but I would argue that we have a natural mission and responsibility to work with them, given the overlap of their interest and expertise with ours, and their potential to generate with us a range of interesting perspectives for others to explore.

Access to data and information

Of the three broad areas of opportunity for on-line users, access to data and information is perhaps the most obvious. It is the logical extension of the CD-ROM-type database to become a multimedia, real-time, networked resource.

The Internet already gives access to a massive resource of scientific information in digital format (which could be described as a de facto distributed science information centre), made up of the millions of Web pages produced by thousands of organisations, groups and individuals. Indexing facilities such as Yahoo and search engines such as Lycos enable one to find information about almost any aspect of contemporary science and technology, even if the quality is variable. When I recently wanted to see for myself the evidence for possible life on Mars, following the analysis of a meteorite, the Science Museum's curator of space science pointed me to a Web address containing the entire, seminal published paper on the topic.[1]

Making use of this vast resource of data and information requires considerable initiative, commitment and existing knowledge on the part of the user. For the more casual on-line visitor it is important to provide more structured routes, and recommended highlights. Those on-line users with explicit educational objectives need a further range of suggestions and support. Much data and information can be presented interactively. The Franklin Institute Science Museum's 'educational hotspots' page holds an interesting list of interactive exhibits, many with a direct relevance to contemporary science.[2] Individual examples include 'The UK's own Bradford Robotic Telescope',[3] available for the public to use over the Internet (there tends to be a waiting list), and the San Francisco Exploratorium's pages on visual and aural perception.[4] The stage beyond this is the development of more complex simulations; an example is the famous virtual frog dissection.[5] With the new software such as Java[6] and Shockwave,[7] the use of video and audio clips, and approaches to virtual reality such as QuickTime VR and VRML[8] (figure 1)—all of which are already in use by science centres and museums—the potential is enormous. As yet, though, we have no idea how large the potential user group might be, nor how much it might pay and for what. Research into the most effective design of presentations is also lacking.

Conferencing and discussion

Videoconferencing

If there is one virtual technology particularly suited to science centres and museums, it

Figure 1. The VRML version of the student union lounge in the Diversity University MOO

must be videoconferencing, as an audiovisual, interactive medium. Imagine a member of staff walking round an exhibition carrying a camera and talking to a group of people in a remote school or community centre. The possibilities for visit planning, follow-up and indeed for special remote events are endless. Videoconferencing is difficult on the Internet at present, because of the problems of low bandwidths and speeds, but the situation is likely to improve. The existing CU-SeeMe software[9] (figure 2) is acceptable for some purposes now over the Internet, and as soon as genuine broadband networks are commonplace the opportunities will be much greater.

We have experimented with ISDN2 systems[10] at the Science Museum. With this sort of bandwidth, a rapidly changing picture, such as would be obtained by walking around with a video camera, is of low quality. In a project with schools in Scotland and in the Southampton area of southern England, scenarios such as a demonstration, interaction with a drama character and a question-and-answer session were set up, with most of the background relatively static. This arrangement allows those parts of the image that do move, particularly the important human face, to be of reasonable quality so one can see the changing detail. The demonstration that was

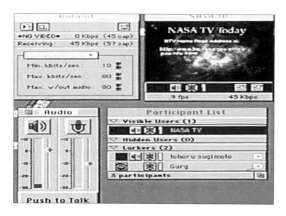

Figure 2. Receiving NASA television via CU-SeeMe videoconferencing

performed over the video link was entitled 'How does a telephone work?'. With close-up shots, it was possible for people at the other end of the link to see all the relevant detail inside the mouthpiece and earphone of a telephone as it was being demonstrated. We have recently installed an ISDN6 system, delivering images of higher quality and allowing the general public and schools to see directly into places of interest such as space centres and computer research laboratories. Contemporary science and technology, and their practitioners, will then become accessible to our visitors in real time.

E-mail conferencing

This may seem like going back in time to a more prosaic and wholly text-based application, because e-mail conferencing has been with us for many years. However, e-mail is far cheaper than videoconferencing, and more controllable. It is possible to collect, read and respond to messages at leisure, allowing more in-depth work.

The Science Museum has run three e-mail conferences to date, on the topics 'Do humans have a role in space?', 'Do we need more roads?', and 'IT—a help or hindrance to society?' Each placed a substantial number of students (about 100) in contact with a range of experts in universities and research centres in the UK and further afield. We are gradually learning how to shape the discussion and make it a genuine dialogue rather than a series of questions and answers. As a means of engaging people in discussion about contemporary science and technology, particularly for schools, this method continues to have much to offer.

Consensus building on the World Wide Web

My third example is an innovative use of the Web. The Science Museum has pioneered the use of consensus conferences in the UK, including hosting the first national consensus conference on plant biotechnology.[11] The concept is that a group of lay people inform themselves about the issues, by reading and by interviewing expert witnesses, and then draft recommendations for future developments. This idea has been extended to exploring the use of the World Wide Web for developing public understanding and the exchange of views on a variety of issues, and for consensus building. Users can read background information as appropriate, and express their views and vote on issues, in ways that immediately affect and update the Web pages themselves.[12]

Collaborative computing

This third type of on-line participation is where I believe that science centres have some radical options for the future. By 'collaborative computing' I mean people working together on-line to construct new electronic resources, including new on-line environments. I have in mind the construction of virtual spaces such as MOOs,[13] with their increasingly graphic capabilities, and shared workspaces.[14] Essentially, one is bringing together the use of the Internet and related systems to provide a massive resource of data and information, with real-time communication between multiple remote users enabling people to construct new cultural resources together.

The electronic science centre of the future

The ease with which Web pages can be created and the democratising (or, for some, anarchic) process that this makes possible means that we now have the opportunity to invite the public at large, expert and non-expert, to help us create on-line science centres and museums. For perhaps the first time in history, the existence of the Internet enables us to put the development of such institutions, apart from their expensive physical realisation, directly into the hands of their communities, rather than building them primarily as the product of a professional management. At the simplest level this can amount to institutions asking participants to submit work, and then selecting and displaying it. A gallery of children's paintings displayed by the Israel Museum is an example of one type of this activity,[15] but I am envisaging something rather more radical, in which real control is placed in the hands of the user.

Science centres and museums that wish to work with and reflect the views of their communities should deliberately encourage a cooperative endeavour with the interested public. This cooperation should take place within a generic but extremely flexible framework, that could rapidly produce a quite staggeringly large, comprehensive and varied series of products and perspectives. Participants would collaborate to construct on-line exhibitions, discuss issues and debate with experts. They would create their own virtual worlds, simulations, libraries and physical exhibits that could be operated remotely. The qualitative difference between this and the development of existing isolated resources by individuals, groups and organisations on the Internet would be that every participant was joining an integrated, cooperative endeavour with the specific aim of creating on-line science centres and museums for themselves and for the interested public at large.

In this way, each science centre and museum would remain a recognised centre of expertise and information, but also become a continuously developing societal construct. Quite apart from the point that I believe we should be doing this anyway, it is not likely to be long before groups come together to construct their own alternative electronic science centres. Individuals are already producing guides to specific museums such as the Louvre[16] and the British Museum.[17] Science centres and museums should lead and work with the process rather than simply react to such future developments.

How realistic is this vision? That depends on how bold science-centre and museum directors are prepared to be, and to what extent these institutions are prepared to surrender some institutional control over content in order to give the public at large more space for expression.

Some facets of this vision are easier to realise in practice. For example, the principle described above can be applied more narrowly to the idea of inviting teachers and students to develop on the Web their own ideas, resources and experiences for using science centres and museums for a physical or on-line visit. The potential scale and richness of the resource that might be created can barely be imagined, and we are now experimenting with this concept at the Science Museum.[18] Given that it is now possible to enable users to annotate Web pages for subsequent readers, and that this may be supplemented by forums (like newsgroups or discussion lists), a rich, professional environment can readily be created.

The Internet is a bottom-up system. No longer is it appropriate, if it ever was, for us to regard ourselves almost exclusively as the experts and the transmission of information as almost invariably one-way: from professionals to visitors. We have to adjust our ways of working—particularly in the electronic domain—and indeed our mind-sets, to release the potential of these media for the benefit of all our on-line users and to create, with interested sections of society at large, the new generation of electronic science centres.

Notes and references

1 http://www.eurekalert.org/E-lert/current/
   public_releases/mars/924/924.html
2 http://sln.fi.edu/tfi/hotlists/interactive.html
3 http://baldrick.eia.brad.ac.uk/rti/
4 http://www.exploratorium.edu/imagery/
   exhibits.html
5 http://www.ncsa.uiuc.edu/SDG/IT94/
   Proceedings/BioChem/robertson/
   robertson.html
6 Java is a programming language developed
   by Sun Microsystems and designed to
   work over networks. It differs from other
   programming languages in that the code is
   transferred from the server and executed
   on the user's computer on the fly. Small
   applications (Java applets) can be written
   using Java to produce animations,
   simulations and interactive Web pages.
7 Macromedia Shockwave is a collection of
   media player software that is installed on
   the user's computer and displays and plays
   back multimedia files, for example audio
   clips, video and small interactive programmes.
8 The Apple QuickTime VR (virtual reality)
   player software is installed on the user's
   computer and plays back three-
   dimensional (3D) images. Images created
   in 3D using special authoring software
   can be manipulated using the mouse and
   the keyboard to create the illusion that
   the user is moving around and
   sometimes through solid objects. VRML
   (virtual reality modelling language) is
   a set of commands that are used to
   describe 3D graphical objects that can
   be viewed, using suitable player software,
   on the World Wide Web. See also
   http://www.ncl.ox.ac.uk/quicktime/
9 CU-SeeMe is an Internet-based videoconferencing program for desktop computers developed by Cornell University's
   Information Technology organisation
   (CIT). Users with CU-SeeMe software
   and an Internet connection can send and
   receive text, audio and video to one
   another or through a central computer
   called a 'reflector'. The reflector supports
   videoconferencing sessions between
   groups of users. Insufficient available
   bandwidth on the Internet or at the user's
   desktop means that video pictures are
   limited in size to 320×240 pixels (about
   the size of a passport photograph) and are
   often received/transmitted at less than
   optimum frame rate. Audio signals compete for bandwidth with text and video
   and so can break up in transmission. At
   present, the text (Chat) window is often
   the best way to communicate.
10 ISDN (integrated services digital network)
   is an international standard for voice,
   video and data digital transmission. ISDN
   allows two or more communication channels to transmit digital data at 64 kilobits
   per second per channel. A separate
   16 Kbps channel can be used for control
   signals and enables a variety of services,
   for example 'call waiting'. Video and multimedia data require large amounts of space
   (bandwidth) for transmission—the more
   ISDN channels that are available the faster
   the data can be transmitted. ISDN2 is the
   most basic service available and provides
   two 64 Kbps channels (a total of 128
   Kbps) over ordinary copper telephone
   wire. ISDN30 and ISDN6 provide 30
   64 Kbps channels (a total of 1920 Kbps)
   and six 64 Kbps channels (a total of 384
   Kbps) respectively over coaxial copper or
   fibreoptic cable.
11 *UK National Consensus Conference on Plant
   Biotechnology, Final Document* (London:
   Science Museum and BBSRC, 1994)
12 http://www.scicomm.org.uk/biosis/human/
   consent.html
13 A MOO is a type of virtual environment
   (domain) that can be accessed by multiple
   users. Users connect to the MOO via the
   Internet and help to build the virtual

environment using very simple object-oriented programming commands. The users communicate with each other within their virtual environment. MUD stands for 'multi-user domain'. MOOs are increasingly being integrated with Web pages, and some are developing VRML environments; http://moo.du.org:8888/

14 http://128.18.101.106:8000/19/vrml10/11/room.wrl; http://bscw.gmd.de/

15 http://www.imj.org.il/3000/gallery/gallerylinked.html

16 http://www.atlcom.net/~psmith/Louvre/

17 http://www.rmplc.co.uk/eduweb/sites/allsouls/bm/ag1.html

18 http://www.nmsi.ac.uk/education/stem/

Turning information into knowledge: the Actua project at newMetropolis

Andrea Bandelli and James Bradburne

The newMetropolis science centre in Amsterdam has developed a strategy for electronic communication to help transform 'visitors' into 'users'.

newMetropolis is Holland's new national science and technology centre, due to open in June 1997. It is located in Amsterdam, in a striking new building (figure 1). At new-Metropolis, the human being is central. Visitors will discover that science and technology stem from human creativity. The offerings are broad: games, experiments, interactive exhibits, demonstrations, workshops and theatre and cinema shows.

One of newMetropolis's goals is to transform our visitors into users, and to help prepare users for the challenges of the next century, when a fundamental skill will be familiarity with global, real-time communication systems, like the Internet. Moreover, newMetropolis aims to garner a reputation for being a source of information about the science and technology that lies behind the news, and for responding to events in the news quickly and effectively by updating exhibitions, programmes and information resources. Part of the way we are meeting this challenge is by developing a computer-based information structure for use by both visitors and staff.

There is now a widespread consensus on the value of the Internet for science centres, by which we mean not only the Internet itself,

Figure 1. An artist's impression of the newMetropolis building, designed by Italian architect Renzo Piano

but all the communication networks that link us to the world and allow us to bring news and information into the science centre at any time. Electronic delivery of information is clearly establishing itself as the way young people prefer to keep updated and to look for information, as witnessed by the constant increase in the number of Internet subscriptions, Internet cafés and information providers.

Science centres and museums cannot remain aloof from this new means of communication for too long. If it is true that the strength of many science museums is in the value of their collections and artefacts, to be fully enjoyed only at first hand, it is equally true that these collections must represent something 'live' and relevant for the young audiences raised in a world where electronic communication is commonplace. Moreover, many science centres do not or cannot rely on artefacts or original collections, but rather on 'experiences' and hands-on exhibits to help their visitors engage with science and technology. In both cases, successful communication can greatly enhance a visit to the museum, not in the limited 'marketing' sense of a one-way message sent to a potential buyer, but as one of the tools a museum can give to its audience—that is the possibility of communicating interactively inside and outside the museum itself.

Transforming visitors into users means giving visitors a reason to return to this institution more frequently than they would normally visit other museums whose collections change infrequently, if at all. To achieve this goal, we have to commit ourselves to continuous renewal and to remaining in touch with the issues that matter to our visitors. Many, if not most, of the issues surrounding science and technology that concern our visitors (our potential users) can be found in the news. Gas attacks in the Tokyo underground, outbreaks of 'mad cow disease' in Britain, bacterial life on Mars: these are issues that spark and catalyse visitors' concerns about science and technology in their own lives. If it is to become a part of people's lives in a meaningful way, the science centre must visibly demonstrate that it is able to respond to issues in the news.

From the outset, we have considered electronic communication not as an end in itself, but rather as the main tool for providing this updated face to our institution. Whether the communication is done over the Internet with a World Wide Web site or with a system of automatic e-mail addresses, is primarily a technical question, not a conceptual one. Our real goal is not just to issue information about our activities, but to build a new tool for the visitors and for the institution itself. This communication infrastructure could easily become part of an exhibition area or an educational project, or become a research facility for our staff. This broad goal to create an interactive information resource can be further divided into two related goals: **1** to update the institution in order to be a centre which is always changing and **2** to create a platform for extension programmes outside the physical site of newMetropolis.

To meet the first of these goals, we need to create a resource within newMetropolis to deal with events in the news, from the latest developments in energy conversion to the global issues of environment, in order to provide updated exhibitions and programmes for the public, and up-to-date research material for staff. Meeting the second goal involves creating a 'virtual museum': a resource available to those who for different reasons cannot come to the physical building. Distant schools and disadvantaged communities are only two examples of large groups that can benefit from the extension programmes made possible by means of electronic networks.

After considering different models of how to use these media in the public and social environments of the museum, we have developed a strategy for electronic communication. Without being exhaustive, our approach groups opportunities into the following four categories: **1** on-floor terminals or public access stations, **2** collaborative education projects with schools, **3** Web sites and remote learning programmes, and **4** the cybercafé.

On-floor terminals or public access stations

For the science centre to become a centre for both internal and external communication, its visitors should have access to the technologies and tools necessary to be in touch with the museum and its staff, and the world at large. However, such an ambition poses a series of questions, for instance: what kind of technologies should be available?, who takes care of the cost of communication?, should the science centre make available the 'knowledge' to communicate or the actual means of communication?, even more elementary, what do we mean by 'communication'?

According to the director of newMetropolis, Joost Douma, science centres should become a 'meeting place' for discussion: a place for visitors, politicians, industrialists and scientists to constantly review and discuss the 'state of our society'. Our exhibitions and programmes all seek to support this goal, and the communication infrastructure we are creating shares the same vision.[1] Communication thus represents the opportunity for our visitors to have access both to the facts as they are presented and to as many points of view on those facts as possible. To these the visitor can add his or her own ideas, opinions and new perspectives, and eventually build up a deeper personal understanding of the subject.

With this in mind, rather than providing public, unlimited access to open networks like the Internet, which risks being confusing and chaotic, we believe it is more worthwhile to give 'glimpses' of what is happening in the world through news broadcasts and to create the possibility of 'exporting' experiences and information learned inside to others. The terminals with open connections to the Internet, which are now beginning to appear in new science centres, are analogous to providing free telephones for the visitors; even if these terminals do partly fulfil the mission of the institution in general, the information they present is completely decontextualised. The alternative is to create an internal information system, a network linked to the exhibitions in the science centre, in order to provide a context in which new information makes sense, and place open access to the Internet where it more properly belongs, in the cybercafé.

Collaborative educational projects with schools

The science centre has a privileged relationship with the school system. It is an informal learning environment and an 'extension' to the existing curriculum. The annual excursion to the museum can be further extended, however, if the science centre can create opportunities to engage the students in long-term learning activities. Electronic networks provide a convenient tool to link schools, especially when they are remote from one another, and to create collaborative educational projects. The science centre then becomes a coordinator for activities that the students perform, a resource to help them 'navigate' in the network of people, institutions and contacts in which the science centre is an important node.[2] The physical collection, the 'museum', may lose some of its importance, but another goal is reached: a science centre made by people, a science centre which represents a living resource for the formal education community. The successful experiments with education projects developed using telematics[3] (for example those developed in Italy by Laboratorio dell'Immaginario Scientifico) are a clear sign that electronic communication is becoming an important tool for the partnership between science centres and schools.

Web sites and remote learning programmes

Distance education is becoming a prioritised issue for many education providers and the possibility of reaching remote users through electronic networks gives the science centres a much broader potential public. In so doing, however, it raises a new series of issues about

competence, target publics and new technologies. The Web has considerably lowered
the threshold to multimedia information from
all over the world by means of its user-friendly
interface and powerful search engines.
Museums and science centres are naturally
among the important players in this emerging
field, mainly because of the large amount of
information they potentially store. Nevertheless, it is only now becoming clear who are the
most important users of the Web in science
museums and science centres. Simply putting
on to the Internet information and content
developed for a completely different kind of
function (for example from exhibitions in the
physical building) is not enough.

Apart from marketing purposes, which of
course play a very important part on the Web,
the potential for remote learning, distance
education and outreach programmes is
immense. The effort of developing new content and new models of interaction between
visitors and the science centre, together with
the definition of new activities and targets,
contributes to continuing experimentation and
shapes the concept of the 'virtual museum',
which for too long has been identified only
with computer-generated three-dimensional
graphic recreations of the museum space.

The cybercafé

Internet cafés, places where people can drink a
cappuccino while sending e-mail messages, are
not just trendy places to spend some time in
the evening. We believe the concept of the
cybercafé has revitalised the entire idea of
early twentieth-century cafés. In those days
people went to the café to meet other people,
read their mail, telegrams and newspapers,
discuss the news, publish artistic, literary, or
political magazines and declaim manifestos.
Businessmen found in the cafés the right
atmosphere to meet other associates, and
artists often saw the café as their interface with
society at large. In many cities (London,
Amsterdam, Vienna and Trieste for instance)
we can still breathe in the atmosphere and the

vitality once found in cafés across pre-war
Europe.

Cybercafés serve the same function,
presented in a new, innovative and up-to-date
way. For a science centre, the cybercafé can
be the direct connection into society: a place
apparently independent of the science centre
where a strong emphasis on communication as
a tool to turn information into knowledge is
always present.[4]

Actua at newMetropolis

From these background considerations, we
have developed the Actua, or news concept,
which is our expression of the concept of the
virtual science centre and the use of the
Internet within the institution.

Access to information

The Internet is constantly referred to as an
endless repository of information. Even without considering the problems involved with
the indexing of such information and, as a
consequence, its retrieval, the main problem if
we look at the learning process while using the
Web is the act of browsing itself. We go from
one page to another, following a thread given
by the free association of thoughts and underlined links, but we often completely miss the
context to which the single pages belong. We
can go from a PhD dissertation to a press
release and from a technical manual to somebody's hobby page in a couple of seconds.
The information we get in this way often
oscillates between valuable content and
useless garbage, even if it concerns the same
topic.

Unless we are only looking for pure data,
building a context for the information available on the Internet is very difficult. A big
helping hand can be given by our institutional
focus as the science centre becomes the place
where all the information is once more put
into the context of the exhibitions, programmes, and public activities. Both visitors
and staff should be able to access not only the

Internet but, more importantly, specific subsets of relevant news and information.

Actua Hoeken: the news corners

The 'news corners' are styled on a common feature of Amsterdam cafés—the reading table. In newMetropolis, these tables host networked computer stations that show the face of newMetropolis daily updated, presenting what is happening now in the fields of science, technology, industry and society. At the same time, the terminals serve as orientation tools, focused not narrowly on the contents of newMetropolis but on the context of the learning experience itself; they provide sample 'tours' through the exhibitions and the programmed events, which are themselves linked to news and current developments in research and science as they are presented by the news.

Moreover, using the Internet, it will be possible to call up references in the news or mark that news to be put in an electronic 'knowledge basket' personalised for each visitor. This is information they can send to their own e-mail address. The Internet thus acts as a 'bridge' to export experience (in terms of information, comments, ideas, contacts) developed at newMetropolis to the outside world.

The workflow

The interface between the institution and the outside world is managed by a special team whose task is to balance the information needs from the inside with the information on offer from outside. The team will be responsible for retrieving the latest news and the relevant information to display in the Actua Hoeken. The sources can be news wires, Internet sites, television stations, etc. The team is also responsible for bringing the pulse of real debates and discussions in science into newMetropolis. In this respect, however, the Actua Hoeken is only one end-user of this information; the entire staff can equally profit from this constant 'open window' to information in order to shape the development of new programmes and improve exhibitions.

From the responses of the users of the Actua Hoeken (both as direct comments and as usage statistics), the team can determine the most frequently accessed information and the most important links to the exhibition. From the staff, the team receives direction on how to conduct further research to develop specific background documentation for newMetropolis. Finally, the team is responsible for managing the communication flow between visitors and staff, for example in terms of requests for special programmes and comments on exhibition development.

Networked exhibits

Our collaborative educational projects take the form of network exhibits: structured communication projects in which students and visitors are able to exchange ideas, comments, and opinions. In these exhibits, students from different locations in the world create 'reports' or scenarios on certain topics, which are then made available to visitors on the floor of participating science centres. These exhibits have two main goals: first, to illustrate current social or environmental issues (like sustainable development, recycling, etc.) and foster discussion about them, and second, to use this discussion (in the form of comments, messages, suggestions or debates) as a learning opportunity for the students who can immediately 'test' their ideas with visitors to the institution. In addition, such exhibits have the advantage that they can be shared among different science centres so that every centre has a 'localised' version of the exhibit to serve the practical needs of its own local school community. Thus they represent a collaboration not only between different schools and cultures but between different institutions.

Remote access to the science centre

In the short-to-medium term, we foresee a considerable increase in the number of remote users of science centres. As described above, new content will be developed to suit this new category of users. One field in which we are particularly interested, is using the network to prototype new projects undertaken by the science centre. New exhibits and new strategies for programmes and text labelling, for instance, can increasingly be developed 'virtually' on the Internet. Educational activities can also be supported by remote access to newMetropolis; a customisation of the offer can be easily provided, together with an easy access to the 'knowledge basket' used on the Actua Hoeken.

Conclusion

It is important to emphasise that the key to the survival of our institutions of informal learning, both as institutions and as places, is in meeting the needs of a wide variety of users. We know from the extensive research by Hood into the motivations of our institutions' users[5] that most of our occasional visitors do not come for high-intensity, challenging learning experiences; they come for social interaction in a public setting. However, our frequent users do come for such learning experiences and often leave disappointed. The new media now allow us to provide high-quality learning experiences for these frequent visitors—our real users.

It is interesting to note how the debate within the museum and science-centre community has moved on. Two years ago, at the European Collaborative for Science, Industry and Technology Exhibitions (ECSITE) conference in Amsterdam, it was not unusual to hear museum directors saying, 'I will never allow any kind of electronic communication in my museum; I cannot imagine children forgetting how to use pencil and paper to write to each other.' This cautious approach has been largely overcome, and today we are less worried about the 'danger' of replacing older,

more traditional media or communication channels, but are looking for new opportunities that can be met by the Internet and will be cost-effective.

The concern about money is not an 'absolute' one; even given the budget restrictions we all are familiar with, science museums can, and are willing to, invest in new technologies. Nor are there prejudices about the effectiveness of new technologies when used appropriately. What we are facing now is the need for a more rigorous reflection on the 'missing parts' of communication in science centres. Literature on this subject is extremely scarce and the only reasonable way to explore this field is by means of examples and experimentation. As a consequence, our institutions should take the initiative in developing new products and programmes with new media. Our plans put an important emphasis on electronic media as a primary means of achieving one of our key institutional missions: to transform visitors into users. The Actua project does this both internally, by bringing in information to fuel new exhibitions and programmes, and on the exhibition floor in the form of Actua Hoeken, which provides the visitor with up-to-date background information on events in the news. Most importantly, this is not a one-way, top-down process. By using the electronic media fully, we also create a means for visitors to the institution to create information for others to use, making the science centre both a platform and a forum for public opinion.

As specialists in informal learning, we are well positioned to take a leading role in creating new approaches to this area. We may not be alone in the field, nor should we be, but our future is guaranteed as long as we continue to take the initiative in creating rich informal learning opportunities, inside and outside the institution, in the science centre and on the Internet.

Notes and references

1 Douma, J, *Prototyping for the 21st Century—a Discourse* (Amsterdam: newMetropolis, 1994)

2 Bandelli, A, 'Linking schools through the network', paper to *1994 ECSITE Annual General Meeting* (Amsterdam: IMPULS Science and Technology Center, 1994)

3 Bandelli, A, 'AriaNet—La telematica al servizio dell'ecologia', *LIS Notizie*, 22 (1995), p2; Smaje, H, 'Science goes to school via the super-highway', paper to *1994 ECSITE Annual General Meeting*

4 Pope, I, 'Just what is the future for cybercafés?', *.net*, 2 (1995)

5 Hood, M, 'Leisure preferences are the key to science-centre audience research', paper to *First Science Centre World Congress* (Vantaa: Heureka, 1996)

The virtual visit—virtual benefits?

John Shane

Virtual visits—virtually real?, virtually accessible?, virtually affordable?—virtually useless?

Søren Kierkegaard, Danish philosopher and theologian, once said, 'Life must be lived forwards, though it can only be understood backwards.'[1] That delicate balancing act of living while looking ahead for opportunities and yet understanding by casting occasional glances over one's shoulder is difficult indeed. The difficulty is compounded by the frequent calls to be 'forward thinking' and not 'mired in the past'.

The efforts of museums at presenting science have, in the past, focused on the principles, theories and artefacts of science. Quite recently there has been a trend to shift the focus to the various impacts of science and technology on society, for example the *Science Box* series of exhibitions at the Science Museum, London[2] and the Science Theater programme at the Museum of Science, Boston. This shift is tremendously important because it is much more likely that the impacts of science and technology will provide a 'point of entry' for the general public than will an exposition of, say, the principles of gravity. In other words, as we pursue the goal of engaging the public in science and technology, we are probably on a clearer course if we focus on the ways in which science and technology intersect with people's everyday lives.

The public is certainly not unanimous in its judgments on science. It seems to vacillate between slavish devotion to scientific disciplines and vehement opposition to all or part of their fruits. It is critically important that, as museum professionals, we recognise the significance of the public's ambivalence about science and technology. That attitude flows from very real concerns about what science is all about and whether it—and its handmaiden, technology—are going to save us all from various hideous ends, or whether our efforts in these fields are themselves what are to be feared.[3]

In that spirit, I would like to quote the American nineteenth-century social critic and transcendentalist philosopher, Henry David Thoreau. He had mixed feelings about the general impact of technology in his time. He remarked that 'our inventions are but improved means to an unimproved end', and spoke disparagingly of people who are 'tools of their tools'.[4]

Although Thoreau wrote a century and a half ago, and one could claim that things have changed in that time, we still have important questions to ask about our tools. I am going to ask three questions about information technology that are seldom raised in discussions about the application of technology:

1 Does information technology work?—what is the value that it adds to the experiences our institutions offer?
2 Can we afford it?—do the direct and hidden costs add up to more than its value or potential revenue?
3 Can our audiences use it?—does this technology give us access to all of our audience, or only a portion? And if only a portion, what portion?

Ultimately all of these questions lead to the core question, 'Does information technology deliver on its promise?'[5]

Does information technology work?

At the Museum of Science in Boston, I am in charge of part of a £4 million project

189

involving six museums called the Science Learning Network. Funded by the National Science Foundation and the Unisys Corporation, the project has, as one of its goals, the creation of on-line resources from science museums that will support and stimulate science education. In pursuit of that goal, we have worked hard to evaluate the educational impact of on-line resources. In that context, I have participated actively in trying to answer this first question: 'Does information technology work?'

Whether exploring the literature, or relying on my own experience, I have been able to uncover only doubts and concerns about variously posed questions of efficacy. I have found more questions than answers, and more problems than solutions. Two characteristics of information technology that are often held up as particular strengths of the medium are access to current information and the speed of that access (access in real time). Although those areas do not represent an exhaustive list of the benefits of what might be available on line, they are often enough cited to warrant some exploration.

Access to more information

Proponents of information technology tout increased access to information as one of the technology's strengths. It isn't at all clear, however, that access to information has been a limiting factor in any or all human enterprises for quite a while. Arguably, access to information has not been a problem since the invention of the printing press. In fact, most would argue that there is more information than anyone can satisfactorily use and that the computer has exacerbated the problem.[6] I would further argue that data and information, which computers handle with facility, are quite different from knowledge, or even useful information. Furthermore, technology's facility with data leads to a valuing of data that is all out of proportion to its true usefulness. Access to information is more of a solution looking for a problem than a problem needing to be solved.

Even when we do enter into the vast information-laden cyberspace, we all know that information available on the World Wide Web is unedited and badly catalogued. Finding information is chancy at best; judging its reliability is very difficult. So as well as access, we have to ask what we are being given access to.

Speed of access

'Quicker is better' may be the mantra of contemporary life. The attendant stress on our systems, both physical and technological, is only now being catalogued. But even accepting the premise that fast is good, the reality is that increased popularity of the Internet has resulted in decreased access. Logging in to a popular Web site at peak use time is often a waiting game. Increasing numbers of users combined with demand for more bandwidth to handle complex images, sound and motion is driving response times down and leaving more and more users waiting. So two of the benefits of information technology, access and speed, turn out to be false promises.

Can we afford it?

The second question is the cost of this technology. When we ask, 'Can we afford it?' we have to look at the direct and indirect financial costs, and what Edward Tenner calls the 'revenge of the unintended consequences'.[7]

Information technology is expensive at the outset, and it remains expensive because of the rising curve of expectation that drives rapid obsolescence of hardware and software and keeps most users scrambling to stay up to date. For example, word from the Sloan School of Management at the Massachusetts Institute of Technology is that MBA-degree students now feel that work must be submitted in colour in order to remain competitive. How many organisations have been driven to fancier and fancier desktop publishing, spreadsheet and presentation software by the perceived need to keep up with others?

These are parlous times for museums. The economic realities of our existence must be acknowledged. Information technology is an expensive technology. We must ask hard questions of any new, costly endeavour. Few museums in my experience have budgets that increase regularly or reliably. Many face decreasing or stagnant resources. These realities demand a high level of accountability and clear demonstration of efficacy before we commit new or redistributed budgets. These standards must be applied to the applications of information technology.

Today, most users access the Internet (at least in the US) through a provider which delivers service with a flat monthly fee, and a local phone call which carries no fee. The question of new fees being imposed on users by carriers is newly opened in the US. NYNEX, a large, regional telecommunications company, has proposed charging Internet providers access fees for each call by a subscriber. Their argument is that Internet use of phone lines is much greater on a per-use basis (time and distance) than traditional phone use, so their costs are going up and their system is being overused with no offsetting revenues. Flat monthly charges unrelated to message units or distances will not be possible for long. Although the Internet world and its associated industries are fighting this effort, it is inevitable that steeply increasing costs are on the horizon.[8]

Edward Tenner explores the costs of computer technology that are often overlooked.[9] For instance, he suggests that a study done by Nolan, Norton and Company at such institutions as AT&T, Bell Laboratories, Ford Motor Company, Harvard University and the Xerox Corporation shows that these institutions have added computer workstations without significantly increasing support staff. Although an institution's budget will show expenses of £1300 to £4300 per workstation to include cost of equipment, supplies and in-house technical assistance, the total cost of a single personal computer workstation will be up to £13,000 a year. The difference between direct

costs and the real costs is made up in what the author calls 'peer support'. These are the hours needed to help end-users learn the system and resolve problems. Most of this help is reactive.

Some costs are incalculable—and arguably distant from museums' applications of this technology, but cost is still worth considering in this context because we represent institutions charged with presenting technology as a subject worth studying as well as a means of delivery.

How should we factor-in the risk of addiction for a growing number of computer-obsessed students? In the US, students at several large universities are failing at college as a result of spending too much time logged on to the Web. Sherry Turkle in her book, *Life on the Screen*, describes conversations with 'on-line junkies' in which they admit that they 'don't do RL (real life) as well as VL (virtual life)'.[10] Schools and psychiatric hospitals are setting up counselling centres to deal with this problem and some universities are creating policies that limit a student's time on line.

The other side of the cost equation is the question of revenue that might offset costs. We know that 'RL' visits generate revenues, if there is a fee at the door. But a large part of the reason for the burgeoning growth of the Internet/Web is the low- or flat-cost structure. That situation has puzzled the business community for years. How do you generate revenue from the virtual visit? Commercial success in this environment—other than the service providers and hardware/software companies—is hard to find. Many companies have tried to develop magazines and newspapers on-line, but the revenue picture is not cheery. In the first place, very little has been developed that is not also available in hard copy. Interestingly, the highly touted on-line magazine, *Slate*, came out in a print version less than a year after the electronic version made its debut.

The idea of a computer screen providing an alternative to hard copy has not caught on. When on-line magazines start charging for their services, the involvement of the public generally plummets. Ziff Publications, a publisher of computer magazines, has tried on-line publishing and the results so far have been disappointing, according to its vice president.[11]

To summarise the issues around the question, 'Can we afford it?', I suggest that we must look at direct and indirect costs carefully and then ask whether this effort will create a revenue stream of any size or substance to support the expenses. With the pressure to have access, costs rise. Few, if any, good models for generating revenue from this largely free medium exist. The future must be labelled as 'questionable at best' from a financial standpoint.

Can our audiences use it?

The final question centres on information technology's ability to facilitate communication between museums and their audiences, and the effectiveness of that link.

Ability to facilitate communication

Communication technology is compromised if significant parts of the population have no access to it. Information technology certainly is not broadly available to all of our actual or desired audiences. We are skewing our definition of 'audience' when we impose this technology between 'us' and 'them'. There is great risk of the world being further divided into the 'information haves' and the 'have nots' by an expensive technology that requires other technologies, such as phone lines and electricity, that are not evenly distributed in the world today.[12] As museums seek to reach wider audiences, it is not wise to turn to a technology that could risk making access even more of a problem. Certainly, some of the money spent on communication technology could be better spent on community outreach programmes that are lower tech, lower cost and more effective.

Even for those with current access to the technology there is the struggle to keep up.

For example, an article published in the *Boston Globe* in 1995 described the situation in a Massachusetts district with 700 middle-school students.[13] Only seven classroom computers were available that were not obsolete. In the same article, a high-school technology coordinator is quoted as saying: 'computers, considered cutting edge five years ago, are antiquated now . . . I have a roomful of boat anchors,' referring to the school's 40 Apple IIe which have no on-line capabilities. What is a school system faced with shrinking budgets to do? What choices are we forcing? Should they cut back on teaching staff or other educational resources in order to participate in this technical revolution? The urgency to do so is certainly there. Everyone from the President of the US to the local school superintendent vigorously praises the educational necessity of computers and the information superhighway. But the critical questions of efficacy or trade-offs have been neither asked nor answered.

Effectiveness of on-line experiences

The question of whether information technology is an effective educational tool is simply unanswered. The Internet is not a good tool for finding information. Information on the Internet is unedited, of widely varying quality and often difficult to locate. Furthermore, efforts to evaluate the educational impact of this technology are scanty, at best. Currently, the most common sort of evaluation being done on the Internet is by the use of log files: the simple counting of 'hits'. Even though there is a clear argument for more rigorous and definitive tools, there has been some settling for what is 'doable' and affordable. And in both of those categories, log files seem to be the medium of choice.[14] This expedient solution to evaluation raises questions. Foremost is the question of whether the number of hits is a deep enough measure of goodness. Counting numbers is an inadequate way to answer questions of efficacy of a very expensive technology. We are at risk of developing an evaluation system that has more in common with television rating systems than rigorous evaluation of educational offerings.

Another challenge in the development and evaluation of on-line experiences is that there does not seem to be any way to prototype the materials adequately. Garfinkle and Johnson describe their efforts to use paper models of on-line resources.[15] Generally, the mock-ups are so different from the final product in all relevant dimensions that they are of little use in helping the developers shape the final product. Once a working (on-line) model is built, considerable resources have been expended, and there is general reluctance to scrap the project.

Much more work needs to be done on the actual educational value of the medium before we can be comfortable that the time, energy and money being spent are worthwhile.

We are now arriving at the question of the relationship between the medium and its message, and the fascinating question of how the nature of the Internet/Web alters or shapes communications. Is the computer an isolating medium? Does it lend itself to the kind of group interaction so central to a museum experience? How does it relate to our focus on the 'real'? Can simulations provide genuine experiences that compete favourably with experiencing real objects with other 'real' people? The questions have been asked, but the research has not been done.

The extension is that when we, as museums, tout this source of information (information technology) as a way to learn, we are introducing a bias into the habits of people looking to us as arbiters and purveyors of knowledge. I suggest that we are at risk of diminishing or impoverishing visitor experiences while trying to expand them. At the very least, we need to pursue actively the answers to these very difficult questions. These are questions about the impact of a technology on its users, society and institutions. And these are questions that ought to be an important element of our educational mission as science museums.

In summary, my answers to the three questions I set out at the beginning are as follows. Does it work? Not always. Can we afford it? Probably not. Can our audiences use it? Many cannot. What these lead to is the core question: 'Does information technology deliver on its promise?' Although I've answered many aspects of that question in the negative, there still remains a technological optimism that suggests that all of these problems are but temporary glitches in a rapidly evolving situation.

Some argue that the rapidly changing technology will not get bogged down in its own limitations, but will instead brilliantly leapfrog these bottlenecks with new and amazing feats of technological derring-do. To those holding that point of view I say we are asking the wrong questions and we will inevitably come up with the wrong answers.

The analogy I offer is the fable of our modern highway/automobile-based system of transportation. After the Second World War there was a huge increase in the affordability of automobiles. This led to more automobiles on the roads which led to increases in traffic and congestion, and to housing in suburbs which also led to more traffic. The question asked (which was the wrong question in my opinion) was, 'How can we move this increased load of cars more rapidly?' The answer, which was reasonable considering the question, was wider roads that led to more lanes of traffic that led to superhighways. The result of that line of questioning and answering is evident to all who have driven near cities lately: more traffic and more congestion.

Compare this line of reasoning with the one that suggests that bottlenecks on the Internet will be solved by increasing bandwidth and increasing the speed of the travel of message units. As bandwidth has increased so have both the number of users and the demand for more bandwidth created by the use of sound, motion and more complicated graphics. At the same time there has been no increase in costs to users to modulate access or to underwrite these increasing costs.

What we have here is a modern tragedy of the commons. Like the overgrazed British commons, the Internet becomes a shared resource not protected by shared responsibility. Each user has incentives to maximise his/her 'take' from the commons, and the cost of that increase is perceived to be shared by all users. Therefore the value received falls fully to the individual, but negative impact is seen as shared. The result is the greater and greater use of the commons to the ultimate detriment of all. The reason? The situation was misperceived or, put another way, the wrong questions were being asked.

I hope that I have demonstrated that the right questions are, 'Does information technology work?' 'Can we afford it?, 'Can our audiences use it?,' and finally, 'Does information technology deliver on its promise?'

I would like to end by a paraphrasing of a fairy tale from the preface of *The Cult of*

Information by Theodore Roszak.[16] He refers to the little boy who blurted out the embarrassing truth that the emperor was wearing no clothes. In this modern retelling, the emperor represents information technology. The little boy did not intend to say that the emperor deserved no respect at all. The emperor may indeed 'have had any number of redeeming qualities. In his vanity he had simply weakened to the appeal of an impossible grandeur. His worst failing was that he allowed a few opportunistic culprits to play upon his gullibility and that of his subjects.' Roszak goes on to espouse considerable respect for the computer and its associated technologies, and even a personal fondness for the machine. But, he says, he does want to suggest that the computer, like the too-susceptible emperor, has been 'overdressed in fabulous claims'.

Notes and references

1 Kierkegaard, S, in Hong H V and Hong, E H (eds and translators), *Kierkegaard's Writings: XI. Stages on Life's Way* (Princeton: Princeton University Press, 1989, originally published 1845)
2 Ward, L, this volume, pp83–90
3 Tiles, M and Oberdiek, H, *Living in a Technological Culture: Human Tools and Human Values* (London: Routledge, 1995)
4 Thoreau, H, *Walden, or Life in the Woods* (New York: Houghton Mifflin, 1893)
5 The following references explore the many perspectives which question the real value of information technology in a variety of applications: Anderson, M, 'Perils and pleasures of the virtual museum', *Museum News*, 73 (1994), pp37–38, 64; Hafner, K and Lyon, M, *Where Wizards Stay Up Late: The Origins of the Internet* (New York: Simon and Schuster, 1996); Mintz, A, 'Techno-logic', *Museum News*, 71 (1992), pp44–45; Postman, N, *Technopoly: The Surrender of Culture to Technology*

(New York: Vintage, 1993); Stoll, C, *Silicon Snake Oil* (New York: Doubleday, 1995); Tenner, E, *Why Things Bite Back: Technology and the Revenge of Unintended Consequences* (New York: Knopf, 1996)
6 Roszak, T, *The Cult of Information* (Berkeley: University of California Press, 1986)
7 Tenner, E
8 Lohr, S, 'The great unplugged masses confront the future', *New York Times*, (21 April 1996)
9 Tenner, E
10 Turkle, S, *Life on the Screen: Identity in the Age of the Internet* (New York: Simon and Schuster, 1995)
11 DiPerna, R, personal communication (1996)
12 Mintz, A
13 Hart, J, 'Slogging to get on line', *Boston Globe* (24 October 1995)
14 Bearman, D (ed), *Hands on Hypermedia and Interactivity in Museums* (Pittsburgh: Archives and Museum Informatics, 1995);

Heinecke, A, 'Evaluation of hypermedia systems in museums,' in Bearman, D (ed), *Multimedia Computing and Museums* (Pittsburgh: Archives and Museum Informatics, 1995); Lohr, S

15 Garfinkle, R and Johnson, V, 'Evaluating scientific visualizations', in Bearman, D (ed), *Multimedia Computing and Museums*

16 Roszak, T

Live interpretation of contemporary science and technology

Frank Olsen

The Experimentarium, Denmark, has found that 'Demos on Wheels' are a cost-effective means of enhancing visitor experiences.

Are visitors to the Danish Science Centre Experimentarium interested in exhibitions on contemporary science and technology? The short answer to this is 'no'. The longer answer is 'yes, provided the subject relates to visitors' everyday lives and/or if they have the chance to interact with museum staff'. The lack of interest in science and technology is not confined to our Experimentarium visitors, but affects Danish society in general. Contemporary science and technology is neglected by public-service television, and only the most popular and sensational science magazines such as *Illustreret Videnskab* (Illustrated Science) make significant sales. Few young people in Denmark choose to study natural sciences and engineering. This situation was an important motivation for establishing Experimentarium in 1991, and we believe that we can make a valuable contribution towards getting people—especially young people—to take an interest in science subjects.

By 1996, more than 2 million people had visited Experimentarium, a third of them in school groups. Recently, we carried out a study to find out which areas our visitors thought we could improve.[1] We began by asking whether it bothered our repeat visitors that we renew only about 10 per cent of the exhibition each year. We also asked about scope: which additional topics would our visitors like to see covered? We conducted interviews with visitors as they left Experimentarium. A pattern emerged, and we were able to categorise the advice and criticisms.

We asked 900 visitors to choose 5 out of 16 suggested concerns that the Experimentarium could address. The most popular suggestions were: **1** more places to sit down (44 per cent), **2** a picnic area where visitors could bring their own food (42 per cent), **3** a clearer exhibition layout (27 per cent), **4** more things for children to do (26 per cent), **5** ensure that fewer of the exhibits are out of order (24 per cent), **6** produce smaller, cheaper pocket-guide to the Experimentarium (20 per cent), and **7** introduce more new exhibits (19 per cent).

New exhibits (not just exhibits about contemporary issues) were given a fairly low priority. Museum professionals tend to think of museums and science centres as consisting of exhibitions and programmes, but the museum experience is much more than that: it is also the car park, the toilets, the café, etc. These facilities prove to be at least as important as what is on display.[2]

The meaning of 'contemporary'

What does contemporary science and technology mean anyway? It means different things to researchers, educators and visitors. I would define contemporary science and technology as what is done today. Contemporary science (and applied science) are the areas researched by universities, institutes and industry, e.g. two-dimensional electron gases, scanning atomic force microscopy, ecology and life-cycle analysis. Contemporary technology is the technology we use in our production and service industries, such as microwaves, recycling, optical fibres, biotechnology, car engines, etc. Naturally some of these subjects can be both science and technology, as in the case of biotechnology. There is also a grey area between basic scientific research and research into applied science and technology.

Although the full name of our institution is the Centre for the Presentation of Natural Science and Modern Technology—Experimentarium, we do not boast many exhibits on contemporary science or technology. Most of our 300 exhibits, like those of many other science centres, are on basic phenomena such as sound, light and the human body. We do cover some contemporary issues, such as communication technology and energy technology. We also run workshops and demonstrations on technologies such as making cheese, body lotion and wine gums. The following are two examples of exhibits in which we addressed contemporary issues with both positive and negative results.

The atomic-force microscope

In 1996 we ran a temporary exhibition for two months called *MicroScapes*, for which the atomic-force microscope was the prime focus. The microscope scanned a master compact disc or some diamond dust and displayed the result on a monitor. The exhibition included eye-catching illustrations of interesting scans, posters and labels that told the story behind the microscope and the Danish company that developed it. When the microscope was unmanned, few people paid attention to it, even though it was on a busy visitor route. However, when a guide was operating the microscope it attracted a lot of interest.

The biotechnology debate game

We developed a biotechnology debate game for the European Week for Scientific and Technological Culture in 1994.[3] The game was used at Experimentarium with special groups, but it was not developed specifically for use in science centres. The target audience was 16- to 20-year-olds, and the objective was to generate discussion about the use of genetic engineering in humans and human embryos, and to encourage people to form an opinion. The game requires 20–40 participants and a game leader. The leader divides participants

into groups of five to seven people, introduces the game and the political aspects of the issue, chairs plenary discussions, makes sure the timetable is adhered to and assists groups by clarifying disputed points. The game leader also participates in group discussions by circulating among the groups. By paying particular attention to proceedings in each group, the leader is able to intervene if the debate begins to wane, by suggesting an angle of the issue that the group has not yet considered, and/or encouraging more purposeful debate. A questionnaire is circulated to each group to be completed after the greatest possible consensus has been reached. During the European Week, groups all over Europe played the game and exchanged and compared results. The game has since been adopted in schools in Denmark, where it is widely used. It is attention-grabbing and provokes discussion in a very personal and effective way.

Both these examples illustrate my belief that we have to choose unconventional methods in the science-centre context to treat contemporary issues. The problem is first and foremost to find methods of presentation. Too much of the contemporary is invisible to the naked eye, because either it is too small or it is encased in a computer. Much of it is so sensitive or expensive that it has to be behind glass. Contemporary science is often complex and requires reams of text to explain, or it is so specialised and remote from daily life that it is greeted with indifference by the visitor. The challenge is to find a way to make contemporary science and technology real, accessible and relevant to the visitor. In an ideal world it also has to be interactive and hands-on, because it has to compete with all the other thrilling exhibits.

'Demos on Wheels'

When we opened Experimentarium in 1991, we organised our guides (called 'pilots') along traditional lines: each pilot had a dedicated area with a group of exhibits—a so-called theme island—where they could help the visitors. The interaction between visitors and pilots was not very effective. Both pilots and visitors felt awkward. To overcome this we developed 'Demos on Wheels', movable demonstration stands on laboratory trolleys (figure 1). A number of small experiments are demonstrated. Although they have a sequence, the demonstration is planned in such a way that visitors can join at any time during the 30-minute show. The subject can be anything from the digestive system to navigation to 'polarised art'. There are five demonstration areas in the exhibition hall, and each demonstration relates to the group of exhibits around the demonstration area. Some of the demonstrations include exhibits and some depend solely on the equipment on the trolley. We currently have about 20 different 'Demos on Wheels', and each day we do 5–20 demonstrations, depending on the season. In addition we still have 'traditional' pilots and we also do stage shows.

Figure 1. 'Demos on Wheels' at the Experimentarium

The 'Demos on Wheels' provide an opportunity for pilots to make contact with and talk to visitors. The pilot might stop a passing visitor with the question: 'What do you think happens when I . . . ?' The visitor also feels more comfortable stopping and watching a demonstration: the situation is non-committal and the visitor can walk away at any time. The demonstrations provide opportunities to tell additional or more comprehensive stories about exhibits. Experiments or demonstrations may be carried out using exhibits that the visitors cannot operate themselves for reasons of practicality or safety. Pilots can explain some of the context that can be difficult to communicate using stand-alone exhibits. Information can be tailored to the individual visitor's level of background knowledge and to relate to the individual visitor's everyday experiences. Finally, the visitors can ask questions, which helps them to structure their own learning.

Evaluating 'Demos on Wheels'

Live interpretation is relatively expensive. It is an obvious target for spending cuts, as it is often seen by management as supplementary to the exhibits and a major drain on running costs. We have therefore started to evaluate our 'Demos on Wheels' to measure their effectiveness.[4] We have found that visitors spend a relatively long time at the demonstrations. They thought them interesting and worthwhile the topics were relevant to their everyday experience and 40 per cent of the visitors actively participated in them. A typical comment was: 'The demonstration highlights the exhibits and allows you to understand them better. The dialogue is very important. You can ask questions and the pilot can answer.'

We chose two 'Demos on Wheels' for the evaluation: 'How does a microwave oven work?' and 'What is pressure?' The evaluation was carried out in three sessions over a period of three months, during weekdays and at weekends. We made unobtrusive observations of 313 people who watched a demonstration:

noting how much time they spent there, from where in the exhibition (which theme islands) they had come and what kind of interaction took place. Second, we focused on all visitors at or near the demonstration, counting how many stopped and how many passed by (sample size was 910). Then we conducted 43 interviews with visitors who either watched a demonstration or who passed by without stopping. Our primary goal was to establish, not what visitors had learned, but the level of interest in their experience of 'Demos on Wheels'.

Results of evaluation

Of the 110 visitors who passed close to a 'Demo on Wheels', approximately 18 per cent stopped for more than 5 seconds. The ratios of males to females, and children to adults who watched the demonstration were about equal. On average visitors spent approximately 5 minutes at a 'Demo on Wheels' but as the half-life period[5] is approximately 110 seconds, we found that interested visitors typically spent 10 minutes at the demonstration, but others left rather quickly (within a minute). Adults spent more time at 'Demos on Wheels' than children, and visitors spent more time at 'How does a microwave oven work?' than at 'What is pressure?' The audience for the demonstrations came from all parts of the exhibition, not only from the theme island with which the demonstration was linked.

What was the quality of the interaction? We found that the contact between visitor and pilot was often initiated by the pilot, who attracted the visitor's attention with a question. Almost 40 per cent of the visitors participated actively, and children participated physically more often than adults. Approximately 50 per cent of the visitors at a demonstration were only listening: they kept their distance (more than 1 m) from the laboratory trolley and they did not get actively involved. We categorised these visitors as 'passively listening'. Approximately 10 per cent of visitors, however, were 'actively listening', i.e. they got close to the

laboratory trolley and they changed position to get better views during the experiments. These 'active listeners' did not became active verbally or physically. Approximately 40 per cent of visitors were labelled as 'active' and some were both verbally and physically active. Approximately 35 per cent of these active visitors were 'verbally active', i.e. they became involved in a dialogue with the pilot. Approximately half of active visitors participated in the demonstration, e.g. holding, feeling or measuring something. Approximately 27 per cent of all children watching and 14 per cent of adults became 'physically active'.

In the interviews we found that visitors found the experiments interesting and worthwhile, and furthermore they were able to relate the demonstrations to their everyday experiences. All visitors rated the 'Demos on Wheels' as 'excellent', primarily on the strengths of the pilot and the topic. It was also felt that a good experience at 'Demos on Wheels' was not dependent on active participation. Those who passed by a demonstration said that they did want to see 'Demos on Wheels', just not that particular one at that particular moment, or they said they had not noticed it. All were able to give a definition of 'Demos on Wheels'.

The audiences for the 'Demos on Wheels' were there either by coincidence or because the pilot stopped them with a question. Many visitors said they stopped because they had a microwave oven at home and wanted to know more about it.

Most visitors were able to retell a number of stories or refer to experiments from 'Demos on Wheels'. For instance, they would remember that 'microwaves heat water but oil stays cool', 'the waves are trapped inside the metal cage', 'the water in an egg will boil and make it explode' and 'metal in a microwave oven will generate sparks'. Only a few visitors had more than a superficial understanding of the technical or scientific content of the 'Demos on Wheels'. Almost all of those who emphasised the value of the pilot were adults who had themselves been active and who stayed a long time. All those interviewed said they thought

'Demos on Wheels' could be 20–30 minutes in duration.

Conclusions

The results of this evaluation show that personal contact with members of staff is very important. Other research supports this conclusion. Studies at the Canadian Museum of Civilization have shown that interacting with a host, particularly when the host initiated the contact, was associated with a significantly longer visit (an increase of half an hour).[6]

Live interpretation such as 'Demos on Wheels' presents an excellent way of interesting people in contemporary science and technology. It supplements stand-alone exhibits in that it can communicate context, tailor information to individual visitors' levels of knowledge, and relate to visitors' everyday experiences. Furthermore, it provides visitors with answers to their questions. Live interpretation is costly, but it enhances the visitor's experience not only of the demonstration but also of the whole science centre.

Notes and references

1   Meyhoff, T and Olsen, F, *Projekt Publikum*, internal report (Hellerup: Experimentarium, 1995)
2   Falk, J H and Dierking, L D, *The Museum Experience* (Washington, DC: Whalesback Books, 1992)
3   *European Debate Games on Biotechnology*, European Week for Scientific Culture, Experimentarium, Copenhagen (1994)
4   Meyhoff, T and Olsen, F, *Summative Evaluation af Demoer på hjul*, internal report (Hellerup: Experimentarium, 1995)
5   The half-life period is the time by which half the visitors have left a demonstration.
6   Needham, H, 'Using program evaluation techniques to evaluate live interpretation at the Canadian Museum of Civilization', *Eighth Visitor Studies Conference* (St Paul, USA: 1995)

Engendering equality: a look at the influence of gender on attitudes to science and technology

Sandra Bicknell

There is compelling evidence that gender differences in attitudes to science are socially induced. Museums and science centres should do more to redress the balance.

Museums such as ours have a special obligation to empower women as much as possible in the science and technology fields. The need to do so is apparent.[1]

The subject of this paper—gender influence on attitudes to contemporary science and technology—is one of those annoying areas that, once embarked on, you wish you had never started to explore. It is a field where the more you know the less you understand: 'I thought I knew something about gender. Now I wonder if I know anything!' as one researcher in this field reported.[2] Problems include that definitions of gender and sex are often blurred and obscured, and research is interpreted in a partisan manner with little reflection upon researcher biases and influences on data collection and interpretation.

Throughout this paper I refer to gender differences rather than sex differences because while 'sex' should describe proven biological differences, 'gender' describes a more pervasive and complex social difference influenced by social, political and economic issues. The semantics are further confused by the use of the sex descriptors 'female' and 'male' on questionnaires, as a shorthand for gender roles; the question is not 'are you XX or XY?' but 'do you consider yourself to be female or male?', and the answer is influenced by social factors.

My assessment focuses on six aspects of gender differences and attitudes to contemporary science and technology, and the implications for the presentation of these subjects in museums and science centres. First, I look at whether gender differences exist. Having established that they do, in a broad sense, I then consider whether they are age-related and, if so, at what age they become apparent. I go on to examine why they exist and whether they can be changed, and conclude by suggesting how museums and science centres might begin to address these issues.

Do gender differences exist?

The most compelling general evidence for gender differences in science museum visiting is the stark demographics: male visitors outnumber female visitors. At the Science Museum, London, more than 60 per cent of visitors are male. Similar figures are reported for the National Air and Space Museum in Washington, DC, and the Deutsches Museum in Munich.[3] At the National Railway Museum, York, 70 per cent of visitors are male. Why is this so? Do women choose not to visit these museums? Do women perceive them as irrelevant places that do not reflect their interests? And, what about those women and girls who do visit? Why do they come? Do they visit for themselves or because of other reasons, such as partners or male children?

Consider this anecdotal evidence. In 1990 I ran a focus group with some teenagers. We had been approached by a school's head of science who wanted to do something different with his students. The students were considered to be 'high achievers' and were aged between 13 and 15. This coincided fortuitously with a venture by the Science Museum to gather information to help develop facilities and activities for teenagers. So, we set up a series of activities for the students and their teacher. We asked for a mixed gender group, but the actual group turned out to be all boys. One of my first questions was to ask why there were no girls with them. One muttered, 'That's a good question.' Others offered their views: 'Girls are no good at science', 'They're not interested.' Asked what girls did instead one boy replied, 'They go shopping.'

This example suggests that gender differences in attitudes towards science, and science and technology museums do exist. A review of a range of research work suggests gender differences can also be found in formal education as well as in the work place and generally throughout many societies.

Gender differences and age

Work in the 1980s concluded that 'research has determined that children ignore activities that they feel involved inconsistent gender behaviour'.[4] If this is so, at what age does this become apparent and how far is this true?

There is much anecdotal information, suggesting that gender differences in the use of science centres, and particularly interactive exhibits, suddenly become apparent around the age of seven.[5] Research carried out observing fifth-grade children (10- to 11-year-olds) at a science centre in Canada, suggested that 'boys seemed more "comfortable" with the technology while the girls were less so and appeared inhibited by the boys' presence'.[6] There was however, no significant gender-related difference in the levels of interaction, the number of exhibits visited, or the duration of stay at each exhibit.

An area with a growing body of research is that of computer literacy and computer use. The assumption usually made about computer-operated activities in museums and science centres is that it is the older generation who feel alienated from this technology. There is, however, also a gender-based alienation that can be seen in differential use of computers in schools, computer clubs, at home and in games arcades: it appears that boys have three times as much experience with computers compared with girls.[7]

There are many data to suggest that gender-related choices and preferences about science

and technology are being made by the age of ten—if not before. Work in the formal education arena has shown gender differences in children's attitudes, interests and self-confidence towards science from the age of nine. These differences become more marked with age. Girls develop a more negative attitude towards science and technology and lowering of confidence which is paralleled by a decline in academic performance. This seems to be related in part to the role of teachers in focusing more on boys, as well as disproportionately fewer out-of-school science experiences for girls. Work in museums and science centres supports the observation of greater confidence in science among boys.[8]

Teenagers' interests in science

A commonly held view is that girls favour biology and boys do not, preferring physics and mathematics. A national survey held in 1993 in the United Kingdom suggests that this view is far too simple an interpretation. The survey assessed students' reactions to pairs of abstract statements relating to particular science subjects and statements relating to the application of concepts, for example: 'How do acids and bases react?' [paired with] 'Why is it good that toothpaste is a base, but that apples are acidic?'[9]

This research showed that the most popular topics, for both boys and girls, were biological ones. It also showed that girls were not interested in all aspects of biology and were interested in other science topics. Abstract statements of physical science concepts were unpopular with both boys and girls. There was also some evidence to suggest that girls were more interested in topics that were relevant to social, human or animal needs. This finding is reinforced by work done in the United States in 1987 and 1992. This research involved a survey of 263 students—initially in the seventh grade (13 years old) and in the twelfth grade (18 years) for the follow-up questionnaire. The students were asked for their levels of agreement with 14 statements about animal research. The results showed that the 'greatest distinguishing feature between opponents and supporters of animal research was gender'.[10]

In 1990, a study in New Zealand set out to examine the differences in interest among secondary-school students in applications of school physics. Forty 15- to 16-year-old science students were interviewed and asked to pick, from a list of subjects showing examples of various applications of physics, those that they would be interested in learning about. A further 500 students were asked to complete a questionnaire in which they ranked a list of technological applications according to interest. The findings suggest that, as shown by earlier studies, girls were more interested in medical applications whereas the boys were more interested in the 'technical aspects of devices, especially how they operate and their use of electronics'.[11] It does not necessarily follow that girls are not interested in technology. The point of this brief description is that within a subject there seem to be gender preferences in interests, which again seem to be related to the application of the topic.

One conclusion that can be drawn from the evidence cited above is that it is unwise to make blanket groupings of topics and assume a gender preference. As Qualter said,

boys and girls respond to topics which they see as relevant to their interests; it is . . . what is relevant to them that determines interest rather than some broad categorisation of topics into biological/physical, abstract/application.[12]

Teenagers' attitudes to science

A number of surveys indicate that boys have a more positive view of science than girls, and that boys have a less negative reaction to all types of science and technology than girls. It has been suggested that this might be because of a masculine image of science, and that boys perceive themselves as having a greater likelihood of pursuing a science-related career.[13] Other work in this area has reported that:

there were few gender differences regarding perception of science and scientists; the only exception was that girls were less likely . . . to think that a high degree of intelligence was required for a science career.[14]

This is countered by boys' views that high intelligence was needed for a career in science and that women are not as able in science as men.

In 1989, a study assessed the attitudes towards scientific change of 1800 English and Scottish students aged between 11 and 16.[15] The survey confirmed previous findings of gender differences: 'Males were much more positive about scientific change than females. The more knowledgeable and more educationally successful were more positive about scientific change.'

Another way of looking at potential gender differences and the presentation of contemporary science and technology is to look at children's favourite museums. In 1993, a children's television programme in the UK, *Blue Peter*, asked viewers to nominate the Children's Museum of the Year; some 10,000 children participated. An analysis of the 1866 museums, suggests some gender differences. The top nominations from boys were the Imperial War Museum and the National Railway Museum. The top nominations from girls were the White Cliffs Experience, the Black Country Museum and the Ironbridge Gorge Museum—all of which are contextually rich environments with indoor and outdoor activities, reconstructions and first-person interpretation. The results also showed that 'girls were more likely to nominate a museum for its hands-on opportunities than boys'.[16]

Adults' attitudes to science

A number of evaluations carried out in the Science Museum between 1988 and 1994 asked questions about suggestions for exhibition topics. The assessment was that these responses showed gender differences which suggested that male visitors were more interested in the 'how does it work?' questions and females in 'what does it mean to me?' questions. In what I now believe to be a rather crass and egocentric interpretation I suggested, at the time, that this could be grouped as males being more interested in the concrete and females in the abstract.[17] Other tantalising hints from evaluations of *Science Box*, a series of exhibitions at the Science Museum on contemporary science and technology topics, suggest that it is not that females are uninterested in science, but that the focus is different and relates to more people-centred topics and interpretation: what does it mean to me and other people? Males seemed to be more interested in the control of the technology: how does it work?, what does it do?, I want to understand it (the object itself). A survey conducted in 1994 asked visitors the following: 'Imagine I could tell you anything you wanted to know about this particular object, what sort of questions would you ask?' The responses were categorised by the type of question, for example, 'what is it?', 'how old is it?', 'who invented it?', 'how does it work?', 'what was it used for?', 'who used it?'. Gender differences were minimal except that more males requested information about technical details.[18]

Why do gender differences exist?

Can gender differences be explained by biology, social conditioning or are both responsible? Does social conditioning influence biology? These questions are raised of many issues: from aggression, to sexual orientation, to alcoholism, to intelligence. For some people, gender differences and imbalances have no biological origin, they are wholly derived from cultural factors.

Biology: the nature model

In this model, the theory is simple: there are biological differences between men and women. But the picture is far from clear:

Some authoritative reviews concluded that males are superior on spatial tasks and that females excel on verbal tasks . . . Other, equally authoritative,

reviews conclude that sex differences in cognitive abilities and/or lateralization are minimal and can effectively be ignored.[19]

Motivation is also a relevant concept:

Girls tend to do better than boys on most measures until some time in adolescence . . . The reasons for this poorer performance are still a matter of controversy but . . . it probably involves failure to engage with a difficult subject with a 'masculine' image.[20]

In explaining his theory of multiple intelligences, Gardner makes only one reference to sex differences and that relates to the spatial intelligence, where he speculates that there would be a selective advantage for the male hunter to 'evolve highly developed visual–spatial abilities'.[21] He makes no specific reference to differences between the sexes for linguistic, musical, logical–mathematical, bodily–kinaesthetic, or intra- and inter-personal intelligences.

Cultural factors: the nurture model

Other work identifies four elements that impact on gender differences and science: innate visual–spatial differences between the sexes; gender socialisation at school and home; gender-role stereotypes; and different levels of participation in science both in school and at home.[22] As pointed out in the literature: '[there is] an assumption that males and females are significantly different';[23] 'From an early age, children develop stereotypes that seem to be especially flagrant in the area of sex roles and that prove quite resistant to change';[24] 'Where parents' behaviour is less sex-stereotyped, children's tends to be less stereotyped';[25] and 'In modern societies it is the education system, in conjunction with other social institutions, which helps to perpetuate gender inequalities . . . Schooling, youth cultures, the family and the mass media . . .'.[26]

An example of mass media impact comes from an assessment of the use of science and technology in children's cartoons, showing that most of the characters who used science and technology were male and were usually being aggressive. Female characters, when linked to science and technology activities, were portrayed as able, socially responsible and using science and technology for the good of others.[27]

Visions of the future—ways forward

Discussions of gender differences in science achievement and attitudes towards science have had disappointingly little to suggest in terms of positive action either to accommodate the difference, or to institute change.[28] I suggest three models of ways forward.

Model 1

Don't worry about it at all; the differences are real and correct and there is no need to change what we do. Personally, I find this an unacceptable model—complacency is not a virtue.

Model 2

Accept that there are differences and design gender-focused exhibits, and make some limited attempts to change the balance. The underlying tenet in this approach is that 'difference is no longer equated with inferiority'.[29] The approach requires a willingness to accept that there may be no 'absolute truths to be striven for; knowledge is contextual and pluralistic'.[30] This is contrary, of course, to the approach of many scientists. A model that allows for gender-specific interpretations will need an emphasis on multiple interpretations, an element of ambiguity—this is a risky strategy for science museums with their traditionalist stance of representing science in an authoritarian and factual manner.

However, 'equitable teaching strategies' can have an effect: the use of small groups that work cooperatively, discussions and hands-on activities. 'Pre-existing gender differences in enjoyment of physical science activities can be ameliorated by girls having opportunities to do such activities.'[31] To succeed in exploiting gender differences, exhibits designed for females would include dialogue, debate, working in small groups (possibly in single-sex groups), non-competitive approaches, real-life relevance and a compassionate approach to the impact of science on people and our world (figure 1). Design style also needs to be considered. For example, studies indicate that girls prefer colourful, detailed images of people, plants and animals, and images that are peaceful; boys prefer images that imply action, suspense and danger or rescue, and include vehicles. Some attempts have been made to design computer graphics specifically catering for gender differences.[32]

Figure 1. Exhibits designed to appeal to girls should encourage dialogue, debate and working in small groups

Model 3

Accept that the differences are socially induced and aim to reduce this effect.
The depth of gender reinforcement cannot, in my view, be underestimated. To resist the pressures of prejudice from parents, peers, teachers, cultural institutions (museums, schools, theatre) and the mass media (newspapers, comics, books, television, films) takes some doing. However, this is no reason not to try: 'Sadly . . . neither museum professionals nor the society in general take museums seriously as agents of change.'[33]

This is the most challenging model and one fraught with difficulties. Gender neutrality is not easily achieved.[34] The underlying tenet with this model is that gender differences are more a product of cultural and social interjections than they are of biology. The approach in dealing with this requires the development of exhibits that stem from a philosophy of everyone being different, with different talents and experiences, but where something meaningful can be presented for everyone. Basically, it requires a recognition of the heterogeneity of audiences. The current use of the phrase 'public understanding of science' is both an ambiguous description and a misleadingly simplistic one. What is this homogeneous thing called 'the public'? Does it exist? I suspect not. What is more revealing is an assessment of publics; publics that might be categorised by gender, age, ethnicity, religion, income, educational attainment, interests or backgrounds.

Conclusions

Whether or not one agrees with the need to address the gender imbalances in science, one cannot pretend that they do not exist, at least in the 'western' world. There are similar numbers of female and male students in tertiary education, but there are half the number of females as males studying science and technology (and conversely, there are half the number of male as female arts students).[35]

If, as museum and science-centre professionals, we are to present contemporary science and technology to the public, then we must address present gender differences.

I believe we need to design consciously for different gender interests, attitudes and motivations (model 2)—through the exploitation of multiple meanings and a recognition of the need for multiple outcomes. We cannot pretend that gender differences do not exist—with the exception of very young children. However, there is an even greater challenge: museums and science centres have the opportunity to lead the way in changing public attitudes to science and technology by closing some of the gender gaps (model 3). Is it really acceptable that by the age of seven the social processes of genderising children are already starting to disenfranchise girls from the worlds of science and technology? I think not. And, with the growing importance that science and technology have in contemporary life, why are knowledge and understanding being shared in an inequitable manner?

From a scan of the literature, it is apparent to me that there is a need for evaluative research into gender reactions to contemporary science and technology presentation. If we are to design exhibits that are non-gender-specific, we need to explore more thoroughly what this actually means in terms of content, style of presentation, context, design and use. If we are bold enough to attempt to be agents of change, then we must commission longitudinal research studies to assess effects.

Finally, a word of caution. This paper has given a partial assessment of just one social grouping: gender. It is partial in two senses: it is an incomplete picture of the complexities of gender issues, and it may be a biased view as I cannot deny my own upbringing. It is also possible that to focus on gender differences and the presentation of contemporary science and technology may be a red herring. One study that looked at science achievement and attitudes by gender and ethnicity found that ethnicity was a more significant factor.[36] And here we come back to the issue of who goes and who does not go to museums and science centres, and what are we going to do to redress this bigger issue? Gender differences in attitudes, interest and ability may be less important factors than ones relating to ethnicity and economics. It is also apparent that the differences within the sexes are far greater than differences between them: 'Both women and men span a range of attitudes, from what has come to be known as "techno-fear" to downright enthusiasm.'[37]

In the short term, museums and science centres have the capability to redress at least part of this imbalance, and in the long term they could radically alter all publics'

understandings of science. Contemporary science is clearly an issue that impacts on everyone and 'As [we move] progressively towards . . . a more technological future, it is essential that no group is left behind.'[38]

Acknowledgement

This paper originates from an unpublished presentation by Sandra Bicknell and Claire Seymour entitled, 'Gender differences in attitudes to science', presented at the 1994 seminar *Gender and Science Exhibitions* in London under the auspices of the Science and Industry Curators Group and Women in Heritage and Museums.

Notes and references

1 Crouch, T, 'Learning from the past', in Glaser, J R and Zenetou, A A (eds), *Gender Perspectives: Essays on Women in Museums* (Washington: Smithsonian Institution Press, 1994), p128

2 Cockburn, C and Fürst-Dilic, R (eds), *Bringing Technology Home: Gender and Technology in a Changing Europe* (Buckingham: Open University Press, 1994), p6

3 Thirty-nine per cent of Deutsches Museum visitors are female (Brandener, T, Deutsches Museum, personal correspondence, 1997); and 42 per cent of visitors to the National Air and Space Museum are female (Ziebarth, E K, Smith, S J, Doering, Z D and Pekarik, A J, *Air and Space Encounters: A Report Based on the 1994 National Air and Space Museum Visitor Survey, Report No. 95-4B* (Washington, DC: Smithsonian Institution, 1995).

4 Kremer, K B and Mullins, G W, 'Children's gender behavior at Science Museum exhibits', *Curator*, 35 (1992), p39. This article refers to a number of relevant gender studies, including views from the Children's Museum in Boston which

supports the hypothesis that children under seven show no gender-specific behaviour in terms of their use of exhibits.

5 Kremer, K B and Mullins, G W, p40

6 Carlise, R W, 'What do school children do at a science center?', *Curator*, 28 (1985), pp31–32

7 Wajcman, J, 'Technology as masculine culture', in *The Polity Reader in Gender Studies* (Cambridge: Polity Press, 1994), pp216–25; Baker, D, 'I am what you tell me to be: girls in science and mathematics', in *What Research Says about Learning in Science Museums*, vol. 2 (Washington, DC: Association of Science–Technology Centers, 1994), p31

8 The following references go through these issues in more detail: Greenfield, T A, 'Gender, ethnicity, science achievement, and attitudes', *Journal of Research in Science Teaching*, 33 (1996), pp901–33; Kahle, J B and Rennie, L R, 'Ameliorating gender differences in attitudes about science: a cross-national study', *Journal of Science Education and Technology*, 2 (1993), pp321–34; Meece, J L and

Jones, M G, 'Gender differences in motivation and strategy use in science: are girls rote learners?', *Journal of Research in Science Teaching*, 33 (1996), pp393–406; Carlise, R W.

9 Qualter, A, 'I would like to know more about that: a study of the interest shown by girls and boys in scientific topics', *International Journal of Science Education*, 15 (1993), pp307–17 (quote is from p309)

10 Pifer, L K, 'Adolescents and animal research: stable attitudes or ephemeral opinions?', *Public Understanding of Science*, 3 (1994), pp291–307 (quote is from p291)

11 Jones, A T and Kirk, C M, 'Gender differences in students' interests in applications of school physics', *Physics Education*, 25 (1990), pp308–13 (quote is from p312). This paper referred to two earlier studies which had shown that there was a division of interest according to gender. The first study found that girls were more interested in biological and medical applications whereas boys had greater interest in technological applications. Following up this work, the researchers found that 'boys were more interested in the physical and technological aspects of the world and girls appeared to be more interested in human and natural aspects' (p308).

12 Qualter, A, p315

13 Qualter, A (1993) found that boys had a less negative reaction to all topics than girls and suggested that this might be because of a masculine image of science and boys' perceptions of their likelihood of having a science-related career. Others, such as Baker, D (1994), also refer to these phenomena. Stereotypic views of scientists have been reported in a range of papers and generally conclude with the image of a white-coated, balding white male with spectacles and facial hair—see, for example, Petkova, K and Boyadjieva, P, 'The image of the scientist and its function', *Public Understanding of Science*, 3 (1994), pp215–24.

14 Greenfield, T A, p921

15 Breakwell, G M, 'Gender, parental and peer influences upon science attitudes and technologies', *Public Understanding of Science*, 1 (1992), pp183–97

16 McAuley, S, 'A child's eye view of museums', paper presented at the *Seventh Annual Conference of the Visitor Studies Association* (Raleigh, NC: Visitor Studies Association, 1994)

17 This interpretation is contradicted by others such as Wajcman, J (1994), p221, who has described females as 'only interested in computers as tools for use and application, as "soft masters", as more concrete and co-operative in orientation' with males 'portrayed as fascinated with the machine itself, being the "hard masters" in terms of computer programming, followers of rules and competition'. But, as Wajcman went on to say, these supposed psychological differences have been at least partially discredited along with the views of women being too 'emotional, irrational and illogical, not to mention lacking visual spatial awareness, to be good at mathematics'.

18 Gammon, B, 'What sort of museum objects interest children?', unpublished internal report (London: Science Museum, 1994). Between 1992 and 1995, ten *Science Box* exhibitions, held in the Science Museum, London, were summatively evaluated; the reports are unpublished.

19 Green, S, 'Physiological studies II', in Radford, J, and Govier, E (eds), *A Textbook of Psychology* (London: Routledge, 1991), pp215–16

20 Meadows, S, *The Child as Thinker* (London: Routledge, 1993), p308

21 Gardner, H, *Frames of Mind: The Theory of Multiple Intelligences* (London: Fontana Press, 1993), p184

22 Meece, J L and Jones, M G, p393. Issues of motivational and cognitive processes are also raised, as the purpose of their research was to assess an earlier 'hypothesis that gender differences in science achieve-

ment are due to differences in rote and meaningful learning modes'—girls tending to use the former and boys the latter. Meece and Jones's data did not support this hypothesis. Breakwell, G M (1992) suggests links between 'networks of beliefs about science' and 'gender, school experiences and family expectations'. The research shows the 'vital role of parental attitudes to science in influencing their offspring's attitudes to science'.

23 Meadows, S, *Understanding Child Development* (London: Routledge, 1986), p195

24 Gardner, H, *The Unschooled Mind: How Children Think and How Schools Should Teach* (London: Fontana Press, 1991), p99

25 Meadows, S, p196

26 Wajcman, J, p217

27 Durham, S and Brownlow, S, 'Sex differences in the use of science and technology in children's cartoons', abstract of a paper presented at the *Annual Meeting of the Southeastern Psychological Association* (Norfolk, VA: 1996)

28 However, relevant recommendations are made in Kremer, K B and Mullins, G W, and in Baker, D.

29 Wajcman, J, p221

30 *The Polity Reader in Gender Studies*, p9

31 Kahle, J B and Rennie, L R, p331

32 Rogers, P, 'Girls like color, boys like action? Imagery preferences and gender', *Montessori-Life*, 7 (1995), pp37–40; Jakobsdottir, S and Krey, C, 'Computer graphics: preferences by gender in grades 2, 4 and 6', *Journal of Educational Research*, 88 (1994), pp91–99

33 Sullivan, R, 'Evaluating the ethics and consciences of museums', in Glaser, J R, and Zenetou, A A (eds), p107

34 Cockburn, C and Fürst-Dilic, R, pp16–17, discuss gender neutrality as a male construct and how, as such, it is far from being a value-free issue.

35 Church, J (ed), 'Education', *Social Trends*, 25 (London: HMSO, 1995), p55

36 Greenfield, T A, p929

37 Berg, A-F, 'Technological flexibility: bringing gender into technology (or, was it the other way round?)', in Cockburn, C and Fürst-Dilic, R (eds), p102

38 Greenfield, T A, p929

The attraction of the spectacular

Patrick Besenval

'Spectacular', 'interactive' and 'educational' need not be mutually exclusive concepts. Films can be powerful media for drawing the public.

For ten years I have been responsible for the choice and creation of moving images and projection systems at Futuroscope Park, 380 km south-west of Paris, near Poitiers. The Park is marketed as 'the European Park of the moving image': it is a leisure park which takes as its theme the possible forms of moving images, ranging from giant screens to interactive systems, from three-dimensional (3D) movies to simulators to circular screens.

Futuroscope is not a scientific park, but it contains almost every possible kind of audio-visual technology and projection system. It can, therefore, be seen as a site for experimentation and observation. Also, it may be useful to try and determine which audiovisual systems are the most suitable for scientific presentation, and to compare the perception the public has of each system with the perception we, the decision makers, have of them.

There is a strong belief that 'interactivity is all' when it comes to presenting science. In traditional presentations, as opposed to interactive systems, visitors are seen as passive, and therefore less likely to be receptive to educational material. However, my conclusion based on our experience of interactivity is that the issue is less clear cut.

At La Cité des Sciences et de l'Industrie, Paris, I once observed the behaviour of some young visitors using a computerised device. The exhibit's title was 'Choose your space mission!'—an exciting programme for youngsters. In the beginning, they were very enthusiastic, ready to play and learn together. However, after one or two easy questions, they were asked to choose the type of orbit they would prefer, the kind of engine, and a few other rather more challenging questions. As a result, they made the most typical interactive gesture: they pushed every button then left.

We cannot be content merely with offering interactivity to the public. There is no innate magic in the word itself or in the technology. The visitors' decision to abandon the device was, of course, a way of interacting, but above all it was a way of reacting: in this example it was a way of saying 'no' to the complexity of the lesson and to the questions through which it was being taught. This story is an example of a badly designed interactive exhibit, but the message is not that interactivity itself is bad, but that interactivity is not the dream recipe for scientific presentation that some people think it is.

At Futuroscope I have created an interactive presentation on water and the environment. I tried to keep my La Cité experience in mind, and looked at the reactions of visitors to the other theatres installed in the Park. It was obvious they enjoyed spectacular systems such as IMAX, dome projections, 3D movies, simulators and 360-degree cinema very much. Few people leave these theatres during screenings. Are they passive? No more than you are passive when you read. Have we to stop writing 'passive books' and must we put interactive questions every two pages to be sure readers are learning and paying attention? No, rather, it seems to me that the hidden concept behind passivity as well as activity or interactivity is the concept of involvement. The public, the visitors, must be involved. To get them involved you have to do more than ask them questions. You have to find a way to make them ask questions themselves. If we ask our young visitors to choose the engine of their space shuttle, in order to teach them things about oxygen and combustion in space, we have missed our target. Most of them are not interested in motorisation technology. To give an idea of a possible way of solving this problem, I will describe the configuration of the interactive movie theatre we have developed, patented and hope to sell to science centres.

For the Aquascope (figure 1), we thought it was most efficient to create an environment in the interactive theatre in which couples or teams of two visitors share only one button to select answers. Examples of the questions include 'What is the proportion of water on Earth (40 or 75 per cent)?', 'What is the proportion of water in our bodies?', etc. To decide, visitors have to debate and discuss the issue with their partner and take a position. This is a very strong involving activity. The questions are not posed directly, but they are embodied in two characters in the film. Viewers are asked to decide which character is correct in their opinion. Once you have chosen which of the two characters is right, the correct answer is given and a tune plays which sounds different for right and wrong answers. Your neighbours can hear your tune and you can hear theirs. Everyone knows who was right and who was wrong. Very quickly a kind of competition takes shape between the teams in the theatre. With a question every 45 seconds, the theatre quickly becomes very noisy. People become more and more attentive to the questions they have to answer in order to win. To succeed in getting this response requires a lot of imagination in the design and effort in terms of popularisation when conceiving the scenario and choosing the questions. The developer has to be very clear about the aims of the show and the message to

Figure 1. Aquascope is an interactive film experience that seamlessly blends viewer participation with education and spectacle

be put across using 15 or 17 questions during the 18-minute show.

It is important that the Aquascope is not an overtly educational experience, but a film with a theme of water and what we know about it. There is no divide between a spectacular show and its educational purpose, between seeing a film and being active while looking at it. The recipe is to always mix these ingredients and never forget about the spectator who is entering a park, a museum or a science centre wanting to be reassured that it was right to leave home and come and spend time and money to be there. In no way should the assumption be made that a visitor is primarily expecting to be taught. This is particularly true for the general public, but I think it could be useful to think in the same way for groups, of adults and of children. If visitors feel they could have had the same experience through watching television, reading a newspaper, or through connecting to the Internet—all of which are cheaper, quicker and more easily obtainable—they could feel cheated. All information technologies are in permanent competition with each other, and this has to be kept in mind when conceiving a new offer, a new product or a new show.

So what specifically can a museum, a park or an exhibition offer? They all have in common both a social and collective dimension, and a physical component, which I'll briefly describe.

The difference between self-learning and visiting a science museum is the same as that between watching television at home and going to the movies. One is essentially a solitary activity and the other is a social or collective activity. The attraction of the social activity is why interactive systems, in which we share our experiences and thoughts with others, are more effective, and why audio-visual presentations which resemble movies more than documentaries or docu-dramas are more successful.

A second component is the 'physical' dimension of the show. A spectator is not a 'spirit' going to a museum to be virtually connected with knowledge. A spectator is a human being complete with body, emotions and brain. You have to feed each of these parts. That is why so-called 'special media' are, in my opinion, particularly interesting for presenting science and technology.

After 50 years of television and ten years of computers, we no longer perceive the screen or a monitor as 'an object'. It is a transparent window on to the world and an interface to information. Sitting in front of a 600 m² IMAX screen you cannot forget that the screen is there. You see it, you feel its immensity. Your senses are stimulated by the physical presence of the screen: just seeing it becomes an experience for which you feel justified in having paid money. This is one of the ways that museums can repel competition from any kind of virtual museum. It is possible to link this idea of the physical dimension of a museum, using the different tools special media offer, with new technologies for emphasising the physical dimension of objects.

At Futuroscope I am often asked if new technologies have a great appeal to the public: my answer is generally 'no'. Visitors do not care about the complexity or innovative technology of a device. We usually use technologies as tools, we do not present them as things that have to be shown for themselves. Technology is invisible and fairly uniform. For instance, the same kind of technology is used in a washing machine as in a projection system. There are few entirely 'new' technologies, and even fewer people who know what a specific technology is and how it works. The public interest lies mainly in the effects of technology: the fact that we no longer have to queue to buy a train ticket or to draw money from a bank account. The computer that performs these functions has become as usual and widespread as electric engines. Nevertheless, it would be almost impossible to meet public expectations without these invisible technologies. In general, the success of a park, a museum, a centre, an event or an exhibition depends partly on its novelty and partly on how it is different from other exhibitions or shows.

To be original requires innovation. Let me use the example of a project in the south of France by way of conclusion. A cave, named La Grotte Chauvet, was discovered two years ago. Inside, 30,000-year-old paintings have been found. With the exception of archae-ologists (and myself, because I am working on the project), nobody is allowed to enter the cave so that it remains preserved and research can be carried out. How then could it be possible to offer to the public something that would allow them to see these discoveries? After some research and a lot of thinking I proposed a 3D simulator. Why? Not because it is a new technological presentation or for entertainment, but because the experience of entering a cave is exciting, and for our ancestors it was one of the main reasons for choosing such places as sanctuaries. To sit on a very smooth simulator allows us to recreate their experience and their emotion. Because the relief of the rock was integral in their paintings, I realised that it would be a mistake to view them in the same way as we view decorative canvas paintings. The 70-mm 3D filming system will reinforce the paintings' relief and enable viewers to perceive the importance of three-dimensionality. The simu-lation will open in the year 2000 at Pont d'Arc in the south of France.

I hope the centre will be successful: not because we will have used 'new technology' (the Lumière Brothers invented the 3D cinema in 1903) or created an amusing ride, but because using these technologies gives us a chance to create a centre that could be perceived as contemporary, that respects what there is to show, and that is worth visiting.

Tensions between science and education in museums and elsewhere

Paul Caro

A postmodern approach could be the answer to reconciling the aims of educators, science popularisers and museum professionals in the presentation of contemporary science and technology.

Science museums and science centres are meeting points for science, education and the theme parks of the 'edutainment' era. They are certainly a part of the informal education system. A recent report sponsored by the US National Academy of Sciences concluded that we have to reinvent schools using the tools provided by technology.[1] In this paper I analyse the main components of the education problem to see what practical actions may be suggested to increase knowledge of science and technology, both among the general public and in the classroom for the benefit of young people. I consider three types of tensions. The first is between producers of science and educators. Science is a mushrooming affair which causes scientists to work in very narrow disciplines. They can speak only from a specialist point of view. In contrast, educators want to provide their students with a broad knowledge of the fundamentals of science. The second tension is between the actual topics in the contemporary research fields and the curricula: generally there is no overlap. The third is that there is a gap between the actual teaching of the abstract basis of science according to the curriculum, and the practical and experimental out-of-school use by youngsters of machines with a high technology content.

Science for the scientist

One of the features that characterises contemporary science is the large number of narrow fields into which this body of knowledge has split. For example, in my personal experience (40 years of research at Centre National de la Recherche Scientifique), I have seen what

started out as a very small but homogeneous field (physics and chemistry of lanthanides or rare earths) with a simple goal (to separate the rare earths in a pure state), broken into domains such as solid-state chemistry, metallurgy and spectroscopy. Each of these now has different tools, instruments, theories and practices. The circles of specialised people with whom it is possible to exchange information are now small research groups. Young people may not have the same sense of these irreversible and frequent splittings in science because they are naturally educated within a speciality, with no time to look elsewhere and very rarely the curiosity to do so. Only older scientists might be nostalgic for some former unity.

Topics across disciplines such as crystal-field theory diverged so completely during my career that, as an inorganic-materials spectroscopist, I cannot understand what scientists in that field are doing now for the metallic state, and the reciprocal is also true. I also have the utmost difficulty following what the leading theoreticians (a handful of people) are doing today in advanced atomic spectroscopy (which in fact is now a branch of mathematics, dependent on continuous-group theory). For those unfamiliar with the rare-earths field, it may seem to be merely a sideshow on the science scene. But, across the periodic table of the elements there are 17 rare earths, which is a large proportion of the total, and the application of those elements to technology has transformed our culture and daily life. Colour television screens contain rare-earth phosphors, sound equipment is miniaturised using magnetic rare-earth alloys, medical X-rays are safer, and powerful lasers are feasible thanks to rare earths. People use rare earths every day without being aware of them.

I recall this personal story to stress two points. First, there is a demonstrable continuity between research and the technology behind everyday objects. Second, most people ignore the details of the connection and, to get back to the problem of interpreting science for the public, I am not aware of any science museum that devotes an exhibition to rare earths. The reason for this, I suspect, is that it is evidently a very specialised topic and the educators, the didacticians and the practitioners of pedagogy, want to build exhibitions on the general grounds of basic science. But scientists, on the whole, do not work in basic science; they produce sophisticated science which proceeds from narrow fields. The French formal curriculum not only ignores rare earths (although France leads the world's production of rare earths and it is a popular topic for basic research), but it also ignores the common technologies. French schools describe little, if any, of the technologies with which children are familiar, such as telephones, cars, televisions, radios, etc. Worse still, in schools, 'technology' has broadly come to be considered as a 'parking lot' for those students that the education system believes are unable to cope with the abstractions of formal teaching in the sciences or humanities. 'Le technique' receives a large number of youngsters who are definitely marginalised in the education system. I believe that one of the main components of what is obviously a crisis in French education emanates from a national formal curriculum which has a burden of abstract topics designed to develop the ability to conceptualise. Although this works very well for a minority, it is irrelevant for many, because most children do not see the connections of 'le programme' with what they and their parents experience in real life.

Science for the populariser

I have now been a practitioner of science popularisation for more than 15 years. The recipe I have used for the popular press, radio, television or for public lectures, has always been the same. I select a contemporary science paper from a primary journal (such as *Nature* or *Science*) and I attempt to build a short story from it. I select papers that provide elements from which to build an attractive story: it needs heroes, places, characters and situations which provide the emotional or aesthetic background needed to spur the curiosity of the reader, listener or viewer. The great metaphorical topics, the basic mythologies, are strong motivating forces. For instance, because they provide stories about creation, astrophysics or prehistory are well represented in the media (in contrast to rare earths, however, these fields have not provided anything to change people's lives—they offer only dreams). A list can be made of editorial devices that are likely to attract and hold the attention of audiences.[2] For instance, the description of an amazing catastrophe, such as the Cretacean Yucatan comet impact, is a good example of an advanced science topic surfing on classical mythologies.[3] In the press and on television, there are many stories in which the scientist-as-hero travels across a metaphorical desert (which could be the sea, the Antarctic, the primeval forest, etc.). Sometimes the scientist meets totemic animals (such as whales, bears and dolphins). I believe that science popularisation in the press or on television is not a pedagogical enterprise, but a literary one which handles only those science topics which have some 'romantic' value. This literature is in fact very close to the traditional storytelling of tales, legends and epics. It contributes to culture by bringing new clothes to old themes.

Generally, I would suggest that the main elements in the central body of science are rarely popularised. For instance, chemistry is an area often neglected by the press. This is a difficult discipline to tackle, and a recent investigation shows that only 10 per cent of the US population understands the meaning of the word 'molecule'.[4] Without knowing what 'molecule' means, it is very difficult to talk about chemistry, which leads me back to the educational end of the problem.

There are, however, viewers, listeners or readers who may have a scientific and technical background and who relish information on the cutting edge of research. It may be a small public, a combination of professionals and amateurs, but it does exist. Science popularisation is then a specialised information service targeting a specific audience. However, it cannot be considered to be an educational activity within the framework of a formal curriculum. We were once told by a group of teachers that a collection of short video clips on everyday technologies produced by the Centre National de Documentation Pédagogique was 'very nice', but could only be used in classes within short leisurely breaks because they were not contributing towards the exercises needed to succeed at the exams. This remark clearly suggests education is in opposition to culture in matters of science.

Science for the museum

I have been working at La Cité des Sciences et de l'Industrie since 1989. In the permanent exhibition, *Explora*, and many temporary exhibitions, science and industry are combined to show current and future trends in research, and how they affect the citizen and society. It is hoped that the visitor will grasp what is state-of-the-art in sectors such as visual and audio technology, mathematics and computing. Visitors are also able to get a balanced briefing on contemporary problems, such as those to do with issues surrounding the side-effects of industrial progress, like the environment or energy. Is there any contemporary science in these exhibits? The answer is 'yes'.

The postmodern ideology behind the Cité displays is important. The layout is inspired by the style born of the 1985 Paris exhibition at the Pompidou Centre (Beaubourg) called *Les Immatériaux*, and conceived by the postmodern philosopher Jean-François Lyotard. There is no linearity—visitors are free to organise their visit in whatever way

they please, and the experience will be different for every visitor (figure 1).[5] The exhibitions are organised as a collection of fragments. The choice of the fragments has been made through collaboration between curators and members of the scientific or industrial community. Every fragment represents the partial contribution of someone deeply engaged in research or industry, operating under the constraints and taste of an exhibit designer. The topics are not covered as comprehensively as in a textbook. A collection of points of view, of techniques or of problems is presented, with some structure of course, but with no specific suggestions as to

Figure 1. The exhibitions at La Cité des Sciences et de l'Industrie are organised in a non-linear fashion, allowing visitors the freedom to choose their own routes

which order to follow from one fragment or group of fragments to another. Such a scheme demands effort on behalf of the visitor. They have to attend to the fragments and become involved if they want to glean more than a mere impression. Effort spent investigating the fragments will be rewarded by the discovery of connections with contemporary research and technology (including rare earths), but these will certainly not be directly linked to any formal curriculum. This is because **1** it is highly improbable that a curriculum makes references to advanced practices in science and technology (to give an example: the computer simulation of the growth of a tree), and **2** a postmodern layout, by definition, does not follow the traditional, linear, progressive profile of pedagogy. It offers everything on the same level; it does not establish hierarchies.[6] Postmodernism is currently at the centre of a raging polemical debate on the role of science in society. Some scientists believe that it undermines the scientific way of thinking.[7]

The pedagogical strategy of the Cité

The exhibitions at the Cité are extensively used by school groups, especially the *Classes Villette* events, a week of activities using Cité resources and centred on a technical or scientific theme. School visits are different from those of the opportunistic visitor. Teachers and children do not usually come by chance. Their visit has been prepared beforehand with meetings between Cité staff and teachers, which encourages familiarity with the contents of the exhibits and enables the material on display to be used in an efficient way. This material, being a direct projection from science and industry, demonstrates particular issues within the larger content of the curriculum. Consequently, it provides a chance for discoveries and surprises within a broad scholastic topic, and this sustains the interest of the children.

The pedagogical strategy of the Cité is to use the displayed objects as tools within an interactive framework. They are expected to generate sets of questions. A prominent, visible and recognisable object ('un objet-phare') can be the point of departure. After observation, answers to questions are to be found in several smaller displays. A real presentation (such as the ant-farm in the *Cité des Enfants*) is much more efficient at encouraging active participation than, say, a video. The objets-phares behave in the museums as icons, emblematic pieces connected to science and technology that anyone can recognise as such (e.g. a space rocket, aeroplane, car, skeleton or map).

We do not expect children to master a large number of precise pieces of knowledge during their visit. The important point is to show them a method of collecting information and deriving questions which are sometimes difficult to answer. The Cité staff are not substitutes for teachers, who still have to organise their classes to gather information. For teachers, the main task is to be sure that the method has been understood.[8]

The connections with the curriculum are broad and are left to the discretion of the teachers. The Cité can be used most easily in the framework of multidisciplinary topics linked to societal problems (health versus biology, for instance). Many fragments in the exhibitions can be easily connected to everyday life, social debates and economic, technical and industrial issues.

Other services at the Cité

Weekly conferences are presented on selected topics by French scientists. There are also seminars or colloquia. These provide a direct link between the interested public and the scientific community. The Cité organises debates on issues of contemporary science and technology, many of them in association with other media such as the Parisian daily newspapers. The best-attended debates have been on health issues (such as Alzheimer's disease), or sociological issues (such as the 'parascientific craze'). We have a huge multimedia library, open free of charge to the public, called the Médiathèque, with 300,000 volumes and 4000 scientific documentaries available on line. 'Sciences Actualités' at the Médiathèque, maintained by a team of journalists, presents science and technology news on a day-to-day basis. They also make television documentaries and manage weekly radio broadcasts. The Cité, in addition to housing exhibitions, presents opportunities to meet scientists and to debate science news. The Cité runs a service called 'Science Contact' which aims to bring together journalists and scientists (modelled on the British Media Resources Service). At the Médiathèque, the public has access to files containing information on the 'hot topics' in science. Enquiries, especially from teachers, are dealt with by the staff.

Contemporary science and education, a postmodern solution?

Contemporary science is not presented very well in France. The tension between science and the education system reveals a similar, and rather serious, lack of understanding between scientists and the general public. The image of science is tarnished by its association with the genesis of societal problems, and especially health problems. There is clearly an intellectual anti-science movement going on in the western world.[9] The media give extensive coverage to science, but generally within the 'romantic' framework (with the exception of publications like *La Recherche*). Exhibitions in museums are constrained by the time it takes to interpret a discovery, and by the capacity of their staff to assimilate quickly what is going on in the scientific community. As for education, contemporary research has not penetrated the contents of the formal curricula which strive to provide a comprehensive coverage of the fast progress of knowledge. For many children, science at school is just a boring chore, loaded with formulae and complicated words. If, indeed, we live in a postmodern era, we must understand that encyclopaedism is a dead utopia. It is

impossible to learn everything about all things. Moreover, the ideal of being able to master a few essentials in many domains is no longer feasible: to make a living, very specialised knowledge must now be mastered. And people may have to grasp several unrelated bodies of knowledge in their lifetime.

We may have to learn how to handle knowledge as fragments, but detailed, attractive fragments rather than superficial overviews. A sound way of making progress in that direction might be to start from the contemporary content of primary scientific papers. The subjects dealt with can, in many cases, be explained to the public (as in the tradition of the British weekly magazine *New Scientist*). If the curriculum were to make provision for the inclusion of contemporary science, it may become more meaningful than the current situation in which the curriculum is remote from day-to-day reality.

Interest in an area of science can be aroused in a group of children by formulating an attractive science story selected from any medium. Then, using a hypertextual model like that underlying the World Wide Web, one can proceed in two directions: towards more sophisticated information—the primary sources, or towards more basic explanations, presented in a way that is adapted to the understanding of contemporary science and technology.

Of course, in the process most of the abstract components of the basic topics are lost, especially the formal mathematical apparatus; but presentations should be oriented towards a cultural understanding of the subject. The important pedagogical point is to provide a strategy for establishing meaningful links between fragments. A future curriculum would have the essential task of suggesting starting points for a collective work of research in the classroom. Teachers would have to be trained to use the many components of the informal science teaching systems, including museums. But for France, this would mean the abolition of cherished centralisation, and the liberation of local initiatives.

223

A practical exercise can be conducted on almost any short scientific story in the press or on television. Science-based news stories could be starting points for an exploration of other parts of the formal curriculum. For example, a documentary on the potential for using modern Doppler techniques to observe the movements of the foetus could be a starting point. This allows one to introduce the definition of ultrasound, which may lead on to the basic subject of vibrations (which is in the curriculum). As the Doppler technique is used by bats to echolocate their prey, there is an easy entry to the natural sciences. The Doppler effect itself is experienced by anyone listening to traffic. So different fragments can easily be connected together across a raft of sciences and technologies.

Use of such a method, which is quite easy to apply within the framework of computer-network-assisted education,[10] depends on a change in educational procedures and habits. In a world where adaptation is going to be a basic skill, the students will have to learn how to collect and learn from useful information, even if the subject is at first unfamiliar.

What part can science museums play in an education oriented towards the collecting of fragments of knowledge? They clearly could be resource centres for exhibiting the disciplines underpinning contemporary science and industry. The echography example above may provide a scenario for an exhibition centred on the concept of vibrations in continuous media that would use a number of fragments based on results from contemporary research (non-invasive techniques to see inside the body, the physiology of bats, a network of earthquake detectors, etc.), and interpreted using multimedia.

The nature of the actual topics dealt with by scientists is at the heart of the tensions between education, the scientific community and the public at large. There is a lack of awareness, on the part of the school system and the public, of what science and industry are actually doing today. In my opinion, a practical way to bridge that gap is the technique of linking fragments, as I have described above. Certainly, museums and the other components of the informal education system can work in that direction, but it will require a change in the mind-sets of staff who seem to shy away from the effort and imagination necessary to penetrate contemporary science domains. The scientific community would be in a better position to explain its work if it collaborated with museum staff and educators on exhibitions. Clearly a three-sided task force is needed.

A science museum is part of the urban scenery. It is a public building which is usually rather solemn and impressive in appearance. It has the old function of a temple: it is a repository of knowledge. I believe this function has to be enhanced. The science museum may not be a place to contemplate the greatness of the past, but rather a place in which the contemporary scientific and industrial achievements—and problems—are displayed on public ground. A great didactic effort needs to be undertaken to keep up with the advances of science and industry, but somehow this challenge has to be met if our society is to preserve some unity.

Acknowledgements

The author thanks Mr Jean-Marie Sani of the Direction Jeunesse Formation at La Cité des Sciences et de l'Industrie for helpful information and Dr Roland Jackson of the Science Museum, London, for commenting on the manuscript.

Notes and references

1   A report entitiled *Reinventing Schools, The Technology is Now!* by the National Academy of Sciences and National Academy of Engineering has been on-line since 16 June 1995 on the Internet: http://www.nas.edu/nap/online/techgap/welcome.html

2   Caro, P, 'Science in the medias between knowledge and folklore', in *The Communication of Science to the Public: Science and the Media*, proceedings of the Fifth International Conference in the series *The Future of Science has Begun* (Milan: Fondazione Carlo Erba, 1997), in press

3   Marshall, C R and Ward, P D, 'Sudden and gradual molluscan extinctions in the latest Cretaceous of Western European tethys', *Science*, 274 (1996), pp1360–63

4   Miller, J D, 'Public attitudes and understanding of science and technology', *Science and Engineering Indicators 1996* (Washington, DC: National Science Foundation, 1996)

5   See the articles by Lyotard, J-F, Chaput, T and Delis, P, in *Modernes et Après, Les Immatériaux, Sous la Direction d'Elie Théofilakis* (Paris: Editions Autrement, 1985).

6   Information on postmodernism can be found at http://jefferson.village.Virginia.EDU/pmc/ and http://helios.augustana.edu/~gmb/postmodern/

7   Gross, P R and Levitt, N, *Higher Superstition: The Academic Left and its Quarrels with Science* (Baltimore: Johns Hopkins University Press, 1994)

8   Guichard, J and Sani, J M, *Cahiers Pédagagiques*, 348 (1996), pp58–61

9   Caro, P and Funck-Brentano, J-L, 'L'appareil d'information sur la science et la technique', *Rapport commun Académie des Sciences-CADAS*, 6 (Paris: Lavoisier Tec et Doc, 1996)

10 See for instance http://athena.wednet.edu/index.html; much information about education sites on the Internet may be found at http://www.kn.pacbell.com/wired/bluewebn/ and http://www.mckinley.com/index_bd.html

Science education in European schools and popular culture

Joan Solomon

Differences in cultural traditions in educational thought have profound effects on 'scientific culture' in member states of the European Union.

This paper begins with a short report on the findings of a 1996 project by the European Commission on school science and the future of scientific culture in Europe,[1] and the White Paper that followed it. Then the argument will pick up on the differences between countries concerning the most valued knowledge, education and teaching which constitute the much quoted 'cultural diversity' of the new Europe. I suggest that these rather deep-set cultural differences can also account for some of the differences to be found in the educational programmes offered at interactive science centres.

'Scientific culture' in Europe

The European Commission was interested in the scientific knowledge of the future citizens of Europe for both political and economic reasons. By mentioning 'scientific culture' in the project title, the Commission made clear that the focus of our attention was to be general scientific knowledge in the population and attitudes towards science in the citizens of tomorrow, rather than the educating of research scientists. Calling the commodity 'scientific culture', rather than 'public understanding' or 'scientific literacy', gave the project a slightly different focus. This was not to be a literacy project with the consequent deficit implication, nor a basis for measuring understanding. The ultimate objective was to find a way of weaving the science that children learnt at school so very closely into their daily ways of reacting, thinking, emoting and acting, that it could be incorporated into their familiar thought as the intimate vocabulary of culture.

Research for the Economic and Social Research Council on the informal discussions between 17-year-old students on science-based social issues showed that a common base of scientific understanding was essential for any discussion to take place.[2] Whenever the knowledge was missing or a concept was incomprehensible, talk about the social context, as well as about the scientific content, ceased almost immediately. For personal views and attitudes to be communicated, scientific understanding needs not so much to be conceptually correct as to be comfortably familiar. The European project set out to explore how far, and in what way, school science might lay the basis for such a common scientific culture. The term 'scientific culture' is a much better description than 'public understanding' or 'scientific literacy' of this essential 'taken-for-grantedness' which removes scientific knowledge from its privileged position as expert knowledge and recasts it as something everyone shares. As Clifford Geertz claims in his pithy summary of the latest mode of anthropological investigation: 'We are all natives now!'[3] Scientists and schoolchildren, politicians and parents, hold views common to their groups which can be explored as an anthropologist might explore a tribe living in an alien territory.

School science across Europe

The report on which this paper is based was directed by Professor José Mariano Gago from Lisbon University. The project began by commissioning national reports from every European Union (EU) member country, and from a few outside it, in a common format that would cover: the science content in each National Curriculum; the local and national politics of its delivery; informal education through museums and science centres; the training and status of science teachers; and the reality of equal opportunity for all pupils. The report also contains some comparative figures on educational spending collected from published data, and a summary European Report.

Science itself is a 'knowledge culture', pan-European and international in the old sense of the 'invisible college' which harks back to the scientific renaissance, so one finding was to be expected. Shorn of language descriptors and subject demarcations, the science curricula in secondary schools for students aged from about 11 to 16 were almost identical right across Europe. In this respect cultural diversity was not a problem, at least on paper. Schools try to enculture students into this scholarly milieu, but often fail. A more important point is that the official school curricula of the EU member states never specify the mode of teaching exactly and it may well be this, rather than the subject matter, which most affects students' attitudes towards science.

It soon became clear that 'school science' has quite different knowledge boundaries in different countries. In France, with its high Cartesian tradition, 'science' in school includes mathematics. In Italy, where classics still maintain the cachet they used to have in the UK, science is taught along with business studies and economics, as befits its parvenu status. In Portugal, science has the widest domain of all, including history, geography and the social sciences. Only in the UK, true perhaps to its empiricist traditions, are laboratory investigations mandatory at all ages. These demarcations cannot fail to affect how students see the place of science within knowledge, the strength of its connectedness and its wider significance.

The EU's principle of subsidiarity that it shall only act at levels at which the national government is not the appropriate body brings comfort to minority groups in all countries.

Their existence and identity may depend crucially on the control they are able to exert over the education of their young people and the language in which education is delivered. Many, like the Welsh, simply want to teach in their mother tongue. In the majority of EC member states there are partially or completely separate educational systems, as in Scotland, Belgium and the 16 German *Laender*. In Spain, every province dictates 40 per cent of the curriculum.

New trends in Europe's science teaching

No country stipulates exactly how science should be taught, although official recommendations are often made. In almost every country these include exhortations to teachers **1** to make this content as relevant to the student's everyday life as possible, which must make it accessible to popular culture, and **2** to have their students carry out practical work in the school laboratory which lends a touch of scientific credibility to their activities. Practical work also occupies a special position in understanding the nature of science. Rather than considering it as a way of teaching science, several countries, such as the Netherlands, identify and teach the separate process concepts. Only in the UK is the students' grasp of them assessed by standardised tests at ages 7, 11 and 14.

Science has recently become a compulsory subject in primary schools in all European countries, with the exception of the Republic of Ireland. I believe this to be a most important step forward in giving science a public image of being less 'difficult'. It is hardly possible to believe that what children of five or six learn at school is too hard for most adults to comprehend. In most countries (e.g. Sweden, Spain, Denmark, Portugal, Germany, Greece and Poland) an integrated or 'topic' approach is recommended for the teaching of science at primary level. Often the emphasis is now on 'the environment' (Scandinavia) or on 'nature study' (France). A range of arguments and concerns surround this integrated, or topic-based delivery of primary science. It is often argued that the integrated environmental approach is more appropriate for young children, who learn holistically, and gain a better understanding when knowledge is embedded in everyday contexts. This is obviously important for building a popular scientific culture. It is also thought to be more appealing to the sensitivities of young children and their concerns with animals and plants. It does, therefore, have clear implications for the formation of attitudes towards science. However, this integration may also mean that it is hard to tell exactly what science content is covered. There is quite wide concern (e.g. in Spain, Belgium, Ireland, Sweden and Norway) that the physics and chemistry content, in particular, may have been 'integrated away'.

There is also enormous diversity in the teaching of technology and, indeed, in its very meaning. It is based within the science curriculum at various levels of secondary school in Sweden, Greece, France, Denmark and the Netherlands. In other countries, such as Poland, Spain, Portugal, Belgium and the UK, it is a separate subject in the National Curriculum. In Switzerland, Ireland, Denmark and Germany, technology does not figure at all in the compulsory school curriculum. In Sweden, there has been a recent move away from technology as applied science. As a separate curriculum subject across Europe it now has all three of the following meanings: **1** a study of the main industrial means of production (e.g. Italy) with or without a discussion of social effects and civic implications, **2** information technology and the use of computers, and **3** acquisition of workshop skills for the designing and making of technological artefacts (e.g. the Netherlands, Denmark and the UK at junior secondary level).

The use of stories and controversies from the history of science during teaching is mentioned in some advisory documents, but rarely at the mandatory level. The exceptions to this are Denmark (in physics at higher levels) and the UK at junior secondary level (although rarely observed in practice, according to

school inspectors). The debate on including the history and philosophy of science in science lessons, about which many educators feel strongly, is still unresolved in most countries. England launched its first National Curriculum with a compulsory Attainment Target on 'the nature of science' which included stories from history, but in later versions this has been excluded. In France, the traditional philosophical component in education has been greatly reduced, and that country is now more of a leader in the interactive-science movement than in natural philosophy. However, the evidence on how epistemological messages get through to students depends more on the teacher than on the subject matter in the curriculum.

The history of science adds a human dimension which connects the abstract and mathematical with people's human characteristics. It shows the processes of science and also its labours—the nights of watching lengthening shadows on the moon which preceded Galileo's hypothesis of lunar mountains, or the weeks of anxiety and stress which Pasteur suffered before his 'successes' with rabies vaccination. The human predicament is far easier for young pupils to understand than are mathematical and linguistic concepts. Indeed, the kind of 'mental modelling' to do with what a parent, rather than a mechanical device, might do is an essential life skill of the smallest child as s/he wonders what mummy might say or do if she were there.

Finally, there is great diversity in the education, salary and status of science teachers across the EU. The hidden messages which an ill-paid and harassed science teacher conveys to his/her pupils presents a discouraging picture of what an education in science can offer.

Cultural traditions in educational thought

It is generally accepted that three distinct European objective-systems of education exist, each of which supports what are sometimes called the different 'regulative principles of thought'—humanism, rationalism and naturalism.[4] These traditions help us to under-stand educational differences and similarities in access, the methods of teaching, the status of teachers, attitudes towards assessment and attainment and the relationship between school and community.

Humanism

Humanism focuses upon education for the formation of character. In Sweden and other Scandinavian countries, it is strongly believed that the purpose of education is to develop character and citizenship, just as some schools in Britain in the 'public-school tradition' believe in developing character and leadership. Therein lies a substantial difference. Nor-wegian education favours small village schools which keep in close contact with their commu-nities, while British educational authorities try to close small schools on the grounds of lack of competition. Nevertheless, both countries share a humanist vision of education's mission to cultivate the whole child—moral disposition as well as intelligence, and even bodily well-being, where that is thought to be connected with character. It is a humanistic urge, even though it may appear narrow and chauvinist at times. The kind of education for which they argue can be illustrated by a statement from the English and Wales Board of Education in 1968.

It will be the aim of the School to train children carefully in habits of observation and clear reason-ing, so that they may gain an intelligent acquaint-ance with some of the facts and laws of nature; to arouse in them a living interest in the ideals and achievements of mankind, and to bring them to some familiarity with the literature and history of their own country.[5]

Rationalism

This strand in European education is usually traced back to René Descartes, but it stands equally firmly in the tradition of mediaeval scholasticism. Here it is the intellect of the child which is being trained, and not the

character. It leads to a curriculum of largely abstract topics, especially in mathematics and philosophy, and an assessment method which is severe, rigorous and positivist. Indeed, those in humanist educational systems may look towards the rationalists with some envy. Who can hope to assess the character-forming objectives of a humanist system as rigorously as the logical ones of a rationalist education?

Naturalism

The basic contrast between this tradition and the previous two is that naturalist education is seen as an interior process: a creative one, rather than one received (some of these radical educationalists might well have substituted the word 'inflicted') externally from school teach-ers. From Rousseau to Illich, and William Morris to Pestalozzi and Steiner, there have been voices raised for the past two centuries that questioned whether the school as institu-tion, with its top-down delivery of knowledge, was the right way to educate children. In some countries, such as Denmark, its influence is to be found right at the heart of school and college teaching.

Many of these themes were already to be found in Rousseau's *Emile* first published in 1762. These include respect for the 'natural state' of the child, coupled with observation of natural objects, a deep suspicion of all institu-tions including schools and a dedication to the principle that children should be free to make their own steps forward when they are ready to do so. The educational objectives are free-dom from intellectual shackles and the power of creative thought:

Education comes to us from nature, from men, or from things. The inner growth of our organs is the education of nature, the use we learn to make of this is the education of men, what we gain by our surroundings is the education of things . . . Everything should be brought into harmony with natural tendencies.[6]

Civilised man is born and dies a slave. The infant is bound up in swaddling clothes, the corpse is nailed down in his coffin. All his life long man is imprisoned by our institutions.[7]

Conclusion: adding informal education to the cultural traditions

It was also in the schedule of the European project to report on the use of out-of-school resources. As far as we could see, these varied not only from country to country but also from one locality to another, and no substantial conclusions were drawn. Hence the following remarks stem only from my own experience of children's science centres and museums in the countries I have seen.

I would claim that strong elements of naturalism imbue all our science centres. Most try hard not to provide too much reading material and try to make the 'pilots' or 'explainers' as unlike teachers as possible. It is 'info-tainment' as the jargon has it, attempting not only to provide fun along with the process of learning, but also to play to the children's physical senses, along the lines of the 'natural tendencies' that Rousseau advocated.

Nevertheless, there are other elements about. In Sweden and Finland, the humanism which looks to community involvement is very clear. In Heureka, Finland, local people are encouraged to bring samples of water for analysis. In the UK there are centres, such as VISTA in Faringdon, the Discovery Dome in Edinburgh, and to a lesser extent *Launch Pad* at the Science Museum, London, where children are encouraged to carry out their own experiments. Not only does this illustrate the empirical nature of British science, it also attempts to train the child's character through the traditions of 'scientific humanism' via the disciplines of science, as the scientific humanists of the 1930s advocated:

Because science does not flatter our self-importance, because science makes stringent demands on our willingness to face uncomfortable views about the universe . . .. Social privilege is repelled by (its) mechanistic outlook because of its ethical impartiality . . .[8]

Finally there are to be found, just here and there, the very formal and intellectual science museums which might have warmed the heart of Descartes himself; but I will admit that they are rare and chilly places with many labels, dusty locked show-cases, and formidable worksheets to enforce encyclopaedic learning.

Notes and references

1 Gago, J M, *School Science and the Future of Scientific Culture in Europe. Report to the European Commission* (Brussels: European Union, 1996). For a more accessible and shorter reference see Solomon, J, 'School science and the future of scientific culture in Europe', *Public Understanding of Science*, 5 (1996), pp157–65.

2 Solomon, J, 'The classroom discussion of science-based social issues presented on television: knowledge, attitudes and values,' *International Journal of Science Education*, 14 (1992), pp431–44

3 Geertz, C, *Local Knowledge* (London: Fontana Press, 1993)

4 Fundamental educational analysis has been carried out, notably by Durkheim, E, *The Evolution of European Thought* (London: Routledge, 1977) and McClean, M, *Education Traditions Compared: Content, Teaching and Learning in Industrialised Countries* (London: David Fulton, 1995), but the task here is not so much to reiterate these arguments as to re-examine the basic purposes that education serves.

5 England and Wales Board of Education, 'Elementary Code 1904', in Maclure J S (ed), *Educational Documents: England and Wales 1816–1968* (London: Methuen, 1968)

6   Rousseau, J J, *Emile, or Education* (Letchworth: Dent, 1762, translated 1911), p6

7   Rousseau, J J, p10

8   Hogben, L, quoted in Werskey, G, *The Visible College* (London: Allen Lane, 1978), p112

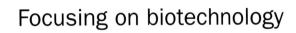

Focusing on biotechnology

Deciding which stories to tell: the challenge of presenting contemporary biotechnology

John Durant

Contemporary science and technology are subject to uncertainty, which means that we have to make radical interpretive choices.

In November 1996, I had the rare privilege of watching a handful of Members of the European Parliament (MEPs) discussing their concerns about biotechnology with a handful of Members of the US Congress. The discussion took place in Washington DC, against the backdrop of growing public protest across Europe about the import of the previous year's US soyabean and maize (corn) crops, both of which contained mixtures of traditional and genetically modified materials.

The MEPs began the discussion by accusing the Members of Congress of having failed in their political duty properly to regulate contemporary biotechnology; and for their part, the Congressmen seemed by turns puzzled and frustrated by their European colleagues' apparent reluctance to accept the obvious benefits of a new and rapidly expanding science-based technology. To those of us who were mere observers at the discussion, it seemed that two rather different political views of contemporary biotechnology prevailed on either side of the Atlantic.[1]

I start with this example of current political debate about biotechnology in order to illustrate a general point that applies to the creation of displays about any genuinely contemporary science or technology, i.e. to any science or technology that is still in the process of coming into being. I'll make this general point straight away and then show how it applies in the case of contemporary biotechnology. Any display or exhibition is necessarily an act of interpretation. Like the author or the broadcaster, the exhibitor is obliged to present some particular view of his or her subject from among a larger set of views that is always available. In principle, any number of alternative narratives may be written; but in practice, only one or at most a few of them must be selected for embodiment (explicitly or implicitly) in an exhibition. What marks out the exhibiting of contemporary science and technology is not the necessity of interpretation, but rather the altogether more radical nature of the interpretive choices that must be made. With the science of the past, we are usually in a position to know what happened—who won or lost any particular scientific dispute, and how this affected the wider society. But in the case of contemporary science, we cannot know these things; and as a result, our interpretive choices are far greater.

Contemporary science and technology are subject to at least two important types of uncertainty. As Simon Schaffer has reminded us, when scientific knowledge is in process of coming into being, the very facts of the matter are commonly in question; and along with the facts, relationships of expertise and authority may also, as he puts it, be 'up for grabs'.[2] This, which may be termed intellectual uncertainty, is commonly combined with ideological uncertainty about the wider implications of the science in question. When scientific knowledge is in process of coming into being, its economic, ethical, industrial, legal, social and political implications may also be in doubt. As a result, the community of those involved in debate about new scientific knowledge is commonly far larger than the network of scientific specialists involved directly in relevant knowledge production or application. Typically, it embraces other scientists or technologists, philosophers, lawyers, journalists and, as my example reveals, policy makers and politicians.

In the case of modern biotechnology, these wider ideological uncertainties are particularly evident. Across the broad field of agricultural, industrial and medical technologies that are currently grouped together under the label of biotechnology, it is easy to identify a long list of substantial issues that continue to be the subject of active public debate. These issues include:

- What are the acceptably safe conditions for the contained use of recombinant DNA technology?
- What are the likely consequences of releasing genetically modified organisms into the environment?
- What are the dietary and health implications of novel genetic foods?
- What regulations should govern the labelling of these foods?
- What are the ethical implications of developing transgenic animals for use as sources of drugs or organs for human transplants?
- What are the medical and moral implications of new genetic tests?
- What, if any, forms of gene therapy are medically and morally acceptable?
- What intellectual property rights should biotechnologists have in their genetic inventions?

It is obvious enough, I hope, that these issues go far beyond the technical uncertainties of contemporary biotechnology. Across the industrialised world, different cultures appear to be responding very differently to particular biotechnologies.[3] In some countries (in particular the US), new biotechnological industries are moving rapidly ahead; while in many others—such as the greater part of the European Union (EU)—rather less progress is being made. In some countries (again the US but also, to some extent, the UK) new agricultural biotechnologies appear to be attracting relatively little public or political concern; while in others (for example, Austria and Germany) they are provoking considerable public resistance. Genetic testing for serious inherited diseases is generally welcomed in some areas; while in others, it is the subject of intense debate. In the US, for example, patents are easily obtainable for biotechnological inventions; while in the EU, for example, they are proving much harder to secure.

These differences cannot be explained in purely technical terms; they are not merely the products of greater or lesser levels of scientific development or technical expertise. Rather, they are the products of particular cultures; of particular historical, social and political circumstances. Praise in one context and protest in another signal different sensibilities, different traditions, different values. Indeed, even the forms of public protest about the same biotechnological issue are often culturally specific. Recently, for example, British journalists have been politely questioning the wisdom of placing genetically modified soya on the food market without labelling. At the same time, Greenpeace supporters in Germany have been dressing up in bunny-suits to protest outside supermarkets and blocking ports to stop genetically modified soya from coming ashore (figure 1); and in Italy, women protesters have stripped naked and 'invaded' Parliament in order to make essentially the same point. These phenomena are not telling us something about biotechnology; rather, they are telling us something about European culture(s).

There is a stark contrast between our comparatively great knowledge about plant, animal and medical biotechnology and our relative ignorance about cultural perceptions of these subjects. However, a certain amount of relevant social research is now under way. For example, the cultural diversity of public responses to biotechnology is the rationale for a major international comparative study of public perceptions that has recently been funded by the European Commission. Research partners in most of the member states of the European Union, Canada and the US are coordinating their efforts to understand public perceptions of biotechnology through three parallel forms of investigation: a study of the development of biotechnology policy over the past 25 years; a study of mass media (especially newspaper) coverage of bio-technology over the same period; and a random sample social survey (conducted in the EU through the *Eurobarometer* instrument of the European Commission).[4] By collecting and collating these three kinds of evidence, the study aims to make sense of the cultural diversity of public perceptions of biotechnology in Europe and North America.[5]

technologies, genuinely interactive exhibits are difficult to devise. These are genuine problems, and I do not mean to minimise them. What I would suggest, though, is that at a deeper level we are faced with even tougher, non-technical problems to do with what I shall term the framing of our subject matter. By 'framing', I mean the ways in which we conceptualise and portray biotechnology itself.

Figure 1. Greenpeace supporters in Germany protest against genetically modified soya

The cultural diversity of public responses to biotechnology has serious implications for those of us who wish to mount public exhibitions on this subject. It is relatively easy to identify the essentially technical problems that confront us in our task: our subject matter is complex and rapidly evolving; much of the relevant science is only partially understood; many of the museological objects that might be used to illustrate the latest developments are hard to find; and other than touch-screen

What is our subject? Why do we select biotechnology (or some other similar construct) as a subject for display, and what do we mean by it? What issues are central to our purpose, and how shall we elect to treat them? In short (since, in my view, narrative is fundamental to the art of exhibiting), what story or stories shall we try to tell? In a situation in which biotechnology is the subject not only of scientific development and commercial application but also of public concern and political debate, we cannot create exhibitions without simultaneously making difficult choices, not merely about the intellectual content but also about the ideological import of our chosen subject matter.

Here, for example, are just a few framing issues to be faced by anyone who sets out to present contemporary biotechnology in exhibitions or displays:

- What is biotechnology? What range of subjects are we including in this construct?
- What is contemporary biotechnology? Is it an evolutionary extension of traditional biotechnology (e.g. of plant and animal breeding), or a revolutionary advance based on gene technology?
- Where is the scientific/technological focus? Are we primarily interested in scientific processes, technological possibilities or industrial products?
- What are the key issues? Are we primarily interested in scientific advances, industrial innovations or ethical/social dilemmas?
- Who are the key figures? To whom do we wish to give a voice?

All media that treat contemporary biotechnology frame it in particular ways. A British newspaper headline that reads 'Crop scientists hit by Greens' action'[6] frames the issue of new food biotechnologies in one way; just as the French newspaper headline that reads, 'Alerte au soja fou' (Alert for mad soya)[7] frames it in another. Museological displays have yet more framing options available to them, not least because exhibitions are particularly good at incorporating and reflecting upon other media.[8]

Our choices, then, are legion; and by the same token, I suggest that our responsibilities as exhibitors are correspondingly great. The question is not merely how to interest our visitors in contemporary biotechnology (though this in itself is extremely important), but with what to engage their attention. Do we wish to arouse their curiosity, to excite their admiration, to provoke their wonder, to inform their understanding, to fuel their concern or to encourage them into some sort of personal or political response? While not mutually incompatible, these are nonetheless different aims; and they are likely to be embodied in rather different types of exhibition and display. In a subject that is beset with so many intellectual and ideological uncertainties, it is surely particularly important that we have a clear idea which stories we want to tell when we set to work to present contemporary biotechnology to the public.

Notes and references

1 The discussion took place in Washington DC on 18 November 1996, during a meeting on 'New Food Products from Biotechnology' organised by the Ceres Forum and co-funded by the US Department of Agriculture and the European Commission.

2 Schaffer, S, this volume, pp31–39

3 Bauer, M (ed), *Resistance to New Technology* (Cambridge: Cambridge University Press, 1995), pp293–331

4 Marlier, E, INRA (Europe) European Coordination Office for the Commission of the European Community DGXII, *Eurobarometer 39.1, Biotechnology and Genetic Engineering: What Europeans Think about It in 1993* (Brussels: Commission of the European Community, 1993). The report is available from the Commission of the European Community, Rue de la Loi 200, 1049 Brussels, Belgium.

5 *Biotechnology and the European Public* is a three-year concerted action plan of the European Commission. It is coordinated by John Durant, Martin Bauer and George Gaskell in London. Further details of the

project may be obtained from: Eleanor Bridgman, Science Museum Library, London SW7 5NH, UK.

6 'Crop scientists hit by Greens' action', *Observer* (15 December 1996)

7 'Alerte au soja fou', *Libération* (1 November 1996)

8 See, for example, Macdonald, S and Silverstone, R, 'Science on display: the representation of scientific controversy in museum exhibitions', *Public Understanding of Science*, 1 (1992), pp69–88.

Dealing with feeling: the challenge of displaying contemporary biotechnology

Robert Bud

Many of the complex and emotive issues of biotechnology can be tackled in exhibitions by presenting original artefacts.

They say miracles are past; and we have our philosophical persons, to make modern and familiar, things supernatural and causeless. Hence is it that we make trifles of terrors, ensconcing ourselves into seeming knowledge when we should submit ourselves to an unknown fear.

William Shakespeare, *All's Well that Ends Well,* Act 2, Scene 3

Since the seventeenth century we have had museums that, too, have made 'modern and familiar, things supernatural and causeless'. It is therefore now a traditional role of these institutions to present new technology to the worried citizen. At the beginning of this century, the Science Museum presented its cutaway Parsons turbine at the same time that turbines were being installed in ships and in the new power stations worldwide.[1] In 1933, a million visitors came to see an exhibition hosted by the Science Museum on plastics, again to introduce a new technology that was making an ever-greater impact on daily life. Such exhibits certainly did explain to visitors how things worked, but that was a secondary objective. They were intended to stimulate thought about the meaning of the objects and of the technology they encompassed.[2] The presence of the artefacts themselves gave them a reality that no other media could bestow. This experience would be repeated in the 1970s when moon-rock was shown to the British public for the first time. Enormous crowds were inspired by the experience of being in the presence of what was a visually unimpressive stone.

For two decades, there has been a sense that biotechnology deserves analogous treatment. The Science Museum has been addressing the problem of how best to represent it since 1980, when the British government's so-called 'Spinks Report'—one of many reports worldwide at the time—urged that the country could not afford to overlook this growth technology.[3] The Science Museum therefore mounted its first exhibition explicitly on biotechnology in 1984.

While much of modern technology is cold and apparently opaque, this subject is hot with emotion-laden characteristics which should render the topic not only important but also engaging to the public. Nonetheless, biotechnology presents distinctive problems to its interpreters. These begin with its very nature. The industry is not yet one of the top-ranked industries in terms of size. Its large number of generally loss-making small companies together enjoy a turnover which may be significant but is hardly gigantic (a survey by *Nature Biotechnology* recently estimated £6 billion for the whole industry, about the same as a single chemical company).[4] Biotechnology is therefore special, and quite different from, say, the turbine industry. It is seen as a technology whose time has not yet come, but is now imminent.

What is 'biotechnology'?

There is enormous public confusion and ignorance about what constitutes biotechnology. This has roots in its diverse, often overlooked history, and is evidenced by the linguistic diversity within Europe.[5] The widely used definition given by the Organisation for Economic Co-operation and Development (OECD) is rooted in a pre-genetic conception, and somehow is remote from the

241

biotechnology which you hear about on the morning news: 'Biotechnology is the application of scientific and engineering principles to the processing of materials by biological agents to provide goods and services.'[6] In German and Danish, there is a distinction between 'Biotechnologie' ('bioteknologi' in Danish) and 'Biotechnik' ('bioteknik' in Danish) which does not exist in English. In Swedish, the word 'bioteknologi' has until recently referred to what in English is called human-factors engineering, while the word 'bioteknik' is used for biotechnology. In the US, the phrase 'genetic engineering' is used synonymously with biotechnology. In Europe, attitudes to the two are rather different. The concept of biotechnology and the use of the word evolved separately among several closed communities and, although there was interbreeding in the 1980s, this has not been sufficient to create a new homogeneous community.

The English word 'biotechnology' has a clear ancestry in the German 'Biotechnologie', which dates back to the First World War and was used in dictionaries in the 1920s. During the early 1970s it was associated with the increasingly significant antibiotics industry, and an important government report identified biotechnology as the interface between microbiology, chemical engineering and biochemistry as a major new industry which would succeed the now mature industries of cars and steel.[7] This report itself spawned the European Federation of Biotechnology, with its still current definitions. Biotechnology grew therefore out of the wish to prolong the 1960s European economic miracle by developing hitherto underutilised but clearly available technologies.

Dimly aware of European excitement over 'biotechnology' and intensely aware of American excitement over the potential of genetic engineering, an American stockbroker used the word biotechnology to describe the activities of young companies which were hoping to exploit the new techniques of recombinant DNA to make valuable proteins.

In the late 1980s, the wish to make commercially valuable proteins using micro-organisms such as *Escherichia coli* was supplemented and even supplanted by the excitement of the Human Genome Project. Now genetic engineering would apply not only to *E. coli* but to humans. Of course the technological and medical objectives were related. The Human Genome Project was quite intentionally developed to stimulate American technology in the face of anticipated Japanese competition, and the techniques and devices developed for sequencing and mapping are technologically important. Moreover, if it were possible to identify the detailed series of chemical transformations linking DNA, RNA, protein and symptom, then one might be able to intervene using a commercial product. However, so far gene therapy has met with little success.

In any case, overshadowing such optimistic, industry-centred visions of biotechnology has been the public reaction. Biotechnology may be defined, not by the excitement it arouses, but by the disquiet. In the 1970s there was anxiety about the escape of dangerous genetic material from laboratories and the potential threats to workers within them. This was succeeded by a more visceral sense of moral danger. The very possibility of modifying genetic inheritance has reawakened fears of eugenics in Germany. The prospect of transgenic animals and the confusion of categories important in our culture has been even more important and perhaps characteristic of biotechnology.[8] Whether we look at the patenting of a mouse, the manufacturing of hormones to increase milk yield from cows, the engineering of a sheep to produce human hormones, or the engineering of a pig's heart to make it suitable for human transplantation, biotechnology is today concerned with the violation of traditional categories which help us live our lives. 'In other words, do you want your milk to come from Daisy or a genetically engineered and manipulated machine?' asked an article in the *Independent* newspaper some years ago.[9] The issue is still relevant, perhaps even more so in the era of bovine spongiform encephalopathy and the evidence that it was caused by the unnatural feeding of meat to cattle.

The chief executive of a major British frozen food company, Iceland, has complained that: 'nature fights back . . . So-called experts allowed dead sheep to be eaten by herbivores and we were used as human guinea pigs. Had the commonsense view of the normal consumers been sought, regardless of scientific evidence, it would not have happened.'[10]

It is possible, therefore, to identify at least five separate interpretations of biotechnology.

- 'Wall Street' definition: what a 'biotechnology company' does
- scientists' definition: applied molecular biology
- journal definition: what is covered by a journal such as *Nature Biotechnology*
- public definition: what people worry about
- historical definition: how biotechnology emerged

A generation ago, the European Economic Community and French delegates to an OECD meeting suggested that the objective of a public information campaign should be 'banalisation'.[11] The public should be reassured that what was at issue was nothing more than the continuation of the tradition of brewing and baking. Today this seems to be quite unfeasible, aside from whether or not it is desirable. In the face of cultural counter-pressures that emphasise the importance and likely radical impact of biotechnology on the public (e.g. through media coverage of dire implications for health insurance, the portrayal of transgenic animals as 'ghastly monsters'), a museum exhibit seeking to counter public fears is likely to be overlooked. At the same time, the hype of millennial technologies is also likely to be misplaced. Change is occurring much more slowly than earlier prophets assumed. In 1996, a quarter of a century after recombinant DNA became a reality, we do not have a single widely used genetic therapy and we have only a few recombinant-DNA-based drugs.

Rather than these extremes that emphasise only outcomes, I would suggest that we focus on the phenomenon itself. Discovery of the changing bounds of tolerance will require engagement with the public. From this perspective, biotechnology is a touchstone of culture as a whole. Understanding something of it is therefore an aspect of public understanding of culture, not just public understanding of science or of industry or even of the future.

Of course, one must address public interest in biotechnology's scientific content. A choice between remarkable outcomes, challenging content, and a developing culture is not appropriate: rather, the intriguing feature of biotechnology is that all these facets are characteristic. In other words, istead of trying to identify the pure strain among these, I would suggest that it is the very diversity of this mongrel that is so interesting.

What and how?

Fundamental to any treatment in the large museum are the questions 'What?' and 'How?'. Is biotechnology a discrete technology which can be dealt with in one place? Or is it better seen as a cluster of approaches to diverse technologies and which should therefore be treated in a decentralised manner in displays on agriculture, chemical industry, medicine and so on? Should biotechnology be equated with genetic engineering of either bacteria or humans? Or is it wider than that? How should we use those typical tools of our trade: the interactive and the historic objects within the displays?

In the context of the large multi-exhibit science museum, it is too much to expect of a single exhibit that it would explore the hopes and fears, scientific workings and technological implications of biotechnology in one space. This also has the danger of isolating the implementation of biotechnology from all the other developments in such fields as pharmacology and agriculture. It would therefore seem to make most sense to treat biotechnology as a cultural phenomenon in one place, and as scientific and technological phenomena elsewhere as appropriate. The Science Museum has sections on bio-technology already within its chemical industry and medical galleries. As we develop new spaces for contemporary science, we will have an opportunity to show more of the context in contemporary research and manufacturing practice.

The response to the question, 'What about biotechnology beyond genetic engineering?' follows on from this. As cultural and scientific phenomenon, the understanding of genetics and recombinant DNA is central. However, industrial proteins and their manufacture, which have their own history, have an impor-tant place. It is an oversimplification to reduce biotechnology to applied genetics, however important and characteristic DNA may be. The issues of making and modifying proteins must have an important role in the culture of biotechnology and should be explored for the public. Moreover, the use of proteins and enzymes, whether in washing powders, mineral extraction or chemical manufacture, has been a distinctive feature of biotechnology as industrial and domestic technology. At the macro-level, the growth of microorganisms is non-trivial and is also familiar to millions of home brewers. Fermentation technology, after all, finds distinctive and economically important expressions in sewage disposal and brewing. Again the decentralised approach to biotechnology would enable us to do due justice to these issues without detracting from the main message about the focus of biotechnology today.

In an area so intrinsically bound up with evolving, and conflicting, cultures, the museum, rather than telling the public what to think, must engage the attention of visitors and stimulate their imagination. Traditionally, there are two powerful tools available to the museum presenter: the historic object and the interactive exhibit. Of course the interactive exhibit is pedagogically powerful and

attractive, particularly to children. However, the cultural implications of new technologies should be the subject of reflection and inspiration, not of pedagogy.

As I have argued at greater length elsewhere, the iconic object is a particularly powerful tool in the presentation of biotechnology.[12] Of course, there are those who just do not respond to the distinctive artefact. Their hearts do not miss a beat when first seeing an historic object, just as some people do not respond to music, while others do. Continuing attendance at museums of all kinds, however, does suggest that there are many who can be engrossed by engagement with the unique artefact. Whether it be an oncomouse with a patent number or an artificial heart, objects that defy conventional categories in particular continue to fascinate.

Emotional power is important, but we must help the visitor to cope with emotion. Classically, objects with such power have been associated with myth.[13] These stories have given context and meaning, helping us to make sense of what otherwise would be threat-ening and haphazard. Myths need to be mean-ingful: their untruthfulness or antiquity is an unnecessary and, in this context, inappropriate adumbration. And our society is littered with such myths about technology, often in specific national contexts. Such myths are particularly important in reassuring us of the categories by which we run our lives and, indeed, stay sane. They are also means of explaining why such categories have changed. 'We discovered penicillin, but the Americans stole it' is a powerful British myth which grew up after the Second World War.[14] Its function was as much to demarcate British identity in the post-imperial age—'poor but creative', and the American identity—'rich but culturally para-sitic', as to make sense of penicillin. The continuing resonance of this myth is shown by the BBC's recent choice of a Science Museum exhibition of penicillin relics to illustrate Britain's continuing inability to build on her people's creativity.[15]

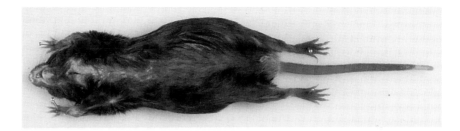

Figure 1. The sexually determined mouse confronts everyone's sense of 'what nature intended'

The Petri dish in which Alexander Fleming first discovered penicillin is thus an object which fits into a complex but meaningful story. The Museum can contrast this with the technology required to turn the 'natural' penicillin into a useful drug and facilitate the public's reflection on the creativity required from technologists and indeed even Americans. The oncomouse, of course, again challenges the line between the 'natural' and the 'technological'. We have displayed it within an exhibition dealing with the chemical industry and near the treatment of quite inanimate machines. The sexually determined mouse developed by scientists at the Imperial Cancer Research Foundation and the Medical Research Council was born female but engineered into maleness.[16] The common response, even from scientists, is one of repulsion. This creature challenges our categories of male and female. It confronts all people's sense of 'what nature intended' (figure 1).

The use of 'monsters' might be alluring but dismissed as facile. What about those impenetrable machines (figure 2)? Here I have to reply: work is in progress. A new exhibit dealing with blood transfusion may not be about biotechnology directly, but it does explore the interpretation of machines. Of all physical materials, blood is perhaps the most confusing: both spiritual material of the highest significance and chemical compound. Many religions give special meaning to the presence or absence of blood. Through expressions such as 'blue-blooded' and 'red-blooded' it is deeply entrenched in our language, yet it can be dried, separated, sold and simulated.

Dr Kim Pelis, a post-doctoral fellow at the Science Museum, is studying the use of objects in the recounting of the history of blood transfusion, and has curated a small temporary exhibition.[17] Here the interplay between blood, technology and life, so vivid in the Dracula story, is explored through the contrast between the very inanimateness of the modern automated blood typing and the animation of our traditional images. As Shakespeare pointed out in *All's Well that Ends Well*:

'Tis only title thou disdain'st in her, the which
I can build up. Strange is it that our bloods,
of colour, weight, and heat, pour'd all together,
would quite confound distinction, yet stand off
in differences so mighty.

The historic object thus has immense power over the imagination and is a particularly appropriate tool in the opening up of issues to do with biotechnology. Exhibits more dedicated to engaging the logical mind are at the same time in danger of losing the sense of drama which is so much associated with

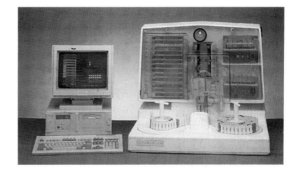

Figure 2. A contemporary blood-grouping analyser developed by Quatro Biosystems

biotechnology for the public. The clarity experienced by the engagement with a video display unit is lost in the outside world when a new challenge is confronted. Understanding biotechnology is more than comprehending the links between DNA, RNA and protein. This does not mean that interactives do not have their place, but rather it places added responsibilities on their designers and on those who would combine their use with that of objects.

So, how are we to answer the challenge of presenting biotechnology? The answer seems to me systemic: combining interactives with objects in different mixes. The benefit might be to engage visitors both with dreams and with realities, so that they can understand the difference between what is likely to be and what might be, what decisions they should be engaged in making and with what dangerous knowledge they will have to learn to live.

Notes and references

1 Dickinson, H W, 'The Science Museum, South Kensington, London', *Mechanical Engineering*, 48 (1926), pp104–08

2 An example of this stimulus was the Festival of Britain. For a reflection of the presentations at the Festival see Cox, I, *The South Bank Exhibition: A Guide to the Story it Tells* (London: HMSO, 1951). See also Weight, R A J, *Pale Stood Albion: The Formulation of English National Identity 1939–56* (University of London, PhD thesis, 1995).

3 ACARD, ABRC and The Royal Society, *Biotechnology: Report of a Joint Working Party* (London: HMSO, 1980)

4 Hodgson, J, 'Companies earn $10 billion of biotech's $30 billion', *Nature Biotechnology*, 14 (1996), pp560–61

5 The etymology of 'biotechnology' is explored in Bud, R, *The Uses of Life: A History of Biotechnology* (Cambridge: Cambridge University Press, 1993).

6 Bull, A, Holt, T G and Lilly, M D, *Biotechnology. International Trends and Perspectives* (Paris: OECD, 1982)

7 Dechema (Deutsche Gesellschaft für chemisches Apparatewesen, chemische Technik und Biotechnologie), *Biotechnologie. Eine Studie über Forschung und Entwicklung—Möglichkeiten, Aufgaben und Schwerpunkte der Förderung* (Frankfurt: January 1974)

8 For an introduction to the importance of categories see Lakoff, G, *Women, Fire and Dangerous Things: What Categories Reveal about the Mind* (Chicago: Chicago University Press, 1990)

9 Blythman, J, 'Will Daisy become a monster?', *Independent* (29 June 1991)

10 'Store chief's fears over designer veg', *Daily Mail* (26 December 1996)

11 OECD, *Biotechnology and the Changing Role of Government* (Paris: OECD, 1988), p72

12 Bud, R, 'Science, meaning and myth in the museum', *Public Understanding of Science*, 4 (1995), pp1–16; also Bud, R, 'The museum, meaning and history: the case of chemistry', in Mauskopf, S (ed), *Chemical Sciences in the Modern World* (Philadelphia: University of Pennsylvania Press, 1993), pp278–94; Bud, R, 'The myth and the machine: seeing science through museum eyes', in Fyfe, G and Law, J (eds), *Sociological Review Monographs 35: Picturing Power: Visual Depictions and Social Relations* (London: Routledge, 1988), pp34–59

13 On mythology more generally see also Northrop, F, *The Great Code: The Bible as Literature* (London: Routledge and Kegan Paul, 1982), pp31–52. For a recent analysis of myth see Liszka, J J, *The Semiotic of Myth: A Critical Study of the Symbol*

(Bloomington: Indiana University Press, 1989). For a new radical sociological version of this appoach see the essays in Mumby, D, *Sage Annual Reviews of Communication Research 21: Narrative and Social Control* (London: Sage, 1993). German literature is reviewed by Jamme, C, *Einführung in die Philosophie des Mythos Vol. 2, Neuzeit und Gegenwart* (Darmstadt: Wissenschaftliche Buchgesellschaft, 1991). The older anthropological literature is reviewed by Gusdorf, G, *Mythe et Métaphysique* (Paris: Flammarion, 1953). Prof Dr Jürgen Teichmann has also examined such an approach: Teichman, J, 'Deutsches Museum: objects, experiment and myth', presentation to the *Workshop on Museums and the History of Technology* (London: Science Museum, 30 July 1996).

14 Wilson, D, *Penicillin in Perspective* (London: Faber and Faber, 1976). See also Bud, R, 'Many happy recoveries', *New Scientist,* 150 (1 June 1996), p48

15 *The Money Programme* transmitted on BBC 2 on 27 October 1996

16 Koopman, P, Gubbay, J, Goodfellow, P and Lovell-Badge, R, 'Male development of chromosomally female mice transgenic for *SRY*', *Nature,* 351 (1991), pp117–21

17 This six-month temporary exhibition opened on 8 November 1996. The Science Museum is grateful to the Wellcome Trust which has funded Dr Kim Pelis's fellowship, jointly held with the Wellcome Institute for the History of Medicine.

Presenting biotechnology: the Heureka experience

Helena von Troil

Biotechnology can be a tricky issue for science centres. The Heureka science centre in Finland has identified the obstacles and sought to overcome them using innovative exhibits.

Modern biotechnology is one of the most important issues in contemporary science, but it is often poorly understood by the public. Sooner or later, everyone will be faced with choices involving biotechnology, such as choosing whether or not to take a genetic test, or selecting between a genetically modified food product and a traditional one. Only with a better understanding of natural sciences in general, and modern biotechnology in particular, can these choices be informed ones. Improving the public understanding of biotechnology is therefore an important task for both scientists and educators. I will discuss four key reasons why biotechnology should be presented in science centres: to provide an open forum for debate, to complement the education system, to showcase national expertise, and to mediate between industry and the public. Then I will show how Heureka, a science centre in Finland, has tackled these issues with reference to specific exhibits.

Providing an open forum for debate

Many applications of modern biotechnology based on genetic engineering have had, or will have, a positive impact on mankind. The rapid growth of this new technology, however, also causes anxieties and doubts about its safe use. In Europe, an open debate on biotechnology is needed in order to define how far we are prepared to go towards changing the genetic make-up of living organisms. It is important to provide lay people with independent, reliable information on the various new applications of biotechnology so that they are better prepared to participate in

this debate. Specialists should engage in the debate and be prepared to address those issues that are relevant to the public.

Complementing the educational system

Science centres are educational institutions whose task is to foster greater public understanding of science and technology. In many places science centres have a strong public-service role. At science centres the visitors are invited to do science themselves. By experimenting and solving problems, they have an opportunity to experience the excitement of science. Interactive exhibits and a relaxed atmosphere facilitate the learning process for both children and adults. Thus science centres are excellent places for presenting themes such as modern biotechnology, where the need to inform and educate the public is great, but the best means to do so are obscure.

Showcasing national expertise

Many science centres collaborate closely with universities and other educational institutions. By doing this they are able to present the national expertise in science and technology. Public awareness of their country's achievements in different areas is, of course, important for the acceptance and support of these institutions. Collaboration with science centres gives scientists an opportunity to interact with the public, explain their work and justify their existence. This is especially important in a field such as genetic engineering, which the general public is hesitant to accept.

Mediation between industry and the public

To the biotechnology industry, public acceptance is crucially important. Not only will the degree of public acceptance influence the behaviour of consumers, but it will also indirectly influence regulation of the use of modern biotechnology and thus the environment in which the industry operates.[1] It is known that 'knowledge, risk perception, and ethical views all influence the degree of acceptability of biotechnology' and that the acceptance also varies considerably according to the application.[2] We also know that optimism with regard to biotechnology is a positive function of objective knowledge of it.[3] For these reasons, it is in the interests of industry that general knowledge of modern biotechnology is improved. Industry, however, is not considered to be a reliable source of information by the public.[4] This makes it difficult for biotechnologists to communicate directly with the public. What industry needs is a mediator, and science centres are good mediators. In collaboration with a science centre, a commercial company has an opportunity to reach an interested and active audience. It should be remembered, however, that the final decision on what to present and how to do it should always be made by the science centre.

European science centres have 20.5 million visitors per year.[5] Presenting biotechnology in science centres will improve the public awareness and understanding of biotechnology and facilitate a better informed, open debate about important issues relating to this technology.

Heureka's exhibition on genetic engineering

Heureka, the Finnish Science Centre, was one of the first science centres to include exhibits on genetic engineering in its permanent exhibition. In January 1993, seven exhibits were inaugurated and they are still on display. The exhibits were, with one exception, financed by Heureka. The theme of genetic engineering was included in the exhibition because at that time applications of genetic engineering were emerging but the general knowledge and understanding of this controversial new technology were poor. In Finland there had been scarcely any public debate about issues relating to modern biotechnology and it was felt that this was an area where Heureka could make a contribution.

To coincide with the opening of this mini-exhibition, public lectures were arranged in the exhibition hall. This initiated the first widespread public discussion about the use of modern biotechnology in Finland. The debate, which became heated at times, continued for several months in all forms of the mass media. This, in turn, led to high numbers of visitors to Heureka. Even college students, who rarely visit science centres, attended the exhibition. The exhibition and the debate surrounding it resulted in scientists and other specialists realising the importance of presenting contemporary science to the public.

Exhibits

The Heureka exhibits can be divided into three groups: those presenting basic facts and principles of genetic engineering; those that present mainly Finnish achievements in this field; and those dealing with ethical issues relating to the use of modern biotechnology. Human-health-related applications were not included because of limited resources.

Two interactive exhibits, 'The Gene' and 'Genetic Engineering and Mutation', explain some of the basic facts needed to enable understanding of what genetic engineering is really about. 'The Gene' explains the function, structure and replication of DNA, and the definition of genes and chromosomes. The visitor is given the opportunity to duplicate a part of the human growth-hormone gene using bases represented by wooden blocks in four different colours.

In 'Genetic Engineering and Mutation' the differences and similarities between these two phenomena are described using a Finnish research project that examined the process by which the colour of flowers can be changed using genetic engineering.[6] This process is compared with mutation by giving examples of naturally occurring plants with flowers of unusual colouring. Here the visitor can take the place of the scientist and deliberately change a flower's colour.

The Human Genome Project is described in an interactive exhibit involving 350 telephone directories. The visitor is asked to decipher a message in DNA code hidden in the page of a telephone book. This exhibit is combined with a short interactive computer program that explains the genetic code and protein synthesis in simple terms. Visitors are asked to build a protein by choosing the right amino acids on the basis of the base sequence in the gene for human growth hormone. Eventually the visitor can build the whole growth-hormone molecule by choosing the correct 191 amino acids.

'The Pigcow' is a visually exciting and controversial exhibit. It presents Finnish and international research projects on transgenic animals. The exhibit is a life-size 'pigcow' (figure 1) connected to an interactive multi-media programme in which the production of modified animals is compared with traditional animal breeding. Transgenic pigs, the production of a transgenic cow, and the oncomouse are presented. This has been the most successful exhibit. It has connections to everyday life, and the scientific content is relevant and interesting. The scientists did not particularly like it as they felt that the exhibit introduced the scientific facts in a controversial way by displaying the pigcow. The media, however, loved it. The creature was used in several television programmes, giving Heureka publicity.

The exhibition also includes two exhibits on the ethics and the morality of the use of genetic engineering. One of these, 'Let's Make a Super Baby' is an audiotaped dramatised argument between two young adults about

Figure 1. 'The Pigcow' exhibit at Heureka. In the background are the telephone directories used to help explain the human genome.

producing a 'super baby' with predefined characteristics. The couple discuss ethical and moral questions in relation to modern methods of in vitro fertilisation. They eventually decide to get a dog instead of having a baby. Ethical problems cannot be solved, but they can be discussed. The discussion is presented on the basis of the ethical conflict model, which also is described.

'What is Your Opinion of Genetic Engineering?' is an interactive computer program that poses a number of questions on the ethics and morality of applications of genetic engineering. The visitor answers questions on genetic testing and the production of genetically modified products and soon finds out that solving one problem leads to another.

Other activities

The opening of the mini-exhibition at Heureka was combined with a series of weekend lectures and discussions on health-related issues, genetic disorders and genetic testing. The lectures were held in the exhibition hall and always started with a brief tour of the exhibits. Visitors were encouraged to ask the lecturer questions. Heureka also invited a drama group from the Stockholm City Theatre in Sweden to perform a play on genetic engineering. The group of three professional actors and one scientist performed their show for school children, politicians and other decision makers in Sweden and other countries. These complementary activities in connection with the exhibition proved very successful in terms of positive feedback and boosting visitor numbers.

Partners

Science centres often produce exhibitions in collaboration with scientific advisors and commercial companies, and Heureka is no exception. Several scientists were involved in the exhibition planning, and efforts were made to collaborate with national biotechnology companies.

Scientists

Specialists in biotechnology are usually scientists, university teachers or product developers from commercial companies, who have their own ideas of what the public 'should know' about genetic engineering. Often, the specialists are not prepared to address the issues that are important to the public because they feel, for example, that these issues are not scientifically correct, they are too emotive or they are based on a complete misunderstanding. However, although the public might have little knowledge or interest in the principles, they are interested in the practical implications of the technology, the global effects on the environment and how the technology influences their life. The scientific experts also tend to overestimate the level of knowledge and understanding of the public. Thus the experts' ideas of what the public wants to know, needs to know and needs to understand may be misguided. This is why the science centre needs to take on the role of mediator.

To create an exhibition on modern biotechnology, it is essential to involve scientists who appreciate the importance of informing the public. They should be included in the team as scientific advisors, but the planning of the exhibition must be carried out by experienced science-centre personnel.

Commercial companies

Although it might be expected that industry would be attracted to presenting new methods and products in a science centre, in my experience this is not the case. Commercial companies are surprisingly sensitive about certain topics and their presentation. For example using the terms 'genetic engineering' and 'genetic modification' may be problematic for some companies, and it discourages their participation in an exhibition on such a theme. Scientists and technologists are often prepared to present biotechnology and the expertise of their company, whereas marketing people are more hesitant. One reason for this might be that there is no consensus within the company on how to deal with the issues of public acceptance. A lack of knowledge within the organisation about biotechnology may be a contributing factor.

Collaboration with industrial conglomerates or associations might be easier than direct collaboration with individual companies. The disadvantage is that decision making is often slow. On the other hand, it has the advantage of avoiding having to coordinate with multiple contacts in many different companies.

Establishing exhibition content

Defining the aim of the exhibition is the first task in the planning of a science-centre exhibition. In the case of presenting contemporary biotechnology, the aim may be to improve the knowledge and understanding of biotechnology among the public. A further aim may be to encourage informed debate about its use or to place certain issues in the arena of public discussion.

Choosing which topics to present can be more difficult than defining the aims, because there are so many important issues relating to biotechnology. The exhibition has to focus on certain well-defined areas, and these areas have to be presented in an attractive, interactive way. The topics should address problems and issues that are relevant to the audience in order for them to bother to interact usefully with the exhibits. For biotechnology this is not a problem, as it is a topic in which both the media and the public are very interested.

So which are the issues in biotechnology that are relevant to the public? According to public-opinion surveys, Europeans are concerned about the safety and ethics of genetic engineering.[7] Lynn Frewer and Richard Shepherd have found that 'ethical consideration plays an important part in the formation of attitudes towards the technology'.[8]

According to the *Eurobarometer* survey, conducted in all the European Union Member States in 1993, research on farm animals, food and plants are the applications of genetic

engineering that have the least public support. Research into medicines and vaccines enjoy more support. Experiments on farm animals and food are the applications that Europeans think are most likely to involve risks to human health and the environment.[9]

From these results we can deduce that exhibits on food, the use of transgenic animals and health-related topics would be of particular interest to the public. The possible risks and ethical and moral aspects of the applications of modern biotechnology should be presented within the displays. Environmental issues also interest the public. With regard to genetic engineering, we know the possible risks to the environment and human health are considered to be important.[10]

Public perceptions are critical for the acceptance of modern biotechnology. But is it the task of the science centre to 'market' this controversial technology? Should we in science centres strive for objectivity or raise questions and challenge the technology? In my view, the facts that we present should, of course, be correct. But by presenting them in different ways it is possible to vary the emphasis and to point out the reasons for controversy.

My experience is that members of the public are more interested than the specialists in controversial exhibits. The most controversial exhibits at Heureka, 'The Pigcow' and 'Let's Make a Super Baby', have been the most successful by far. Both exhibits use art and design to emphasise the controversial nature of the scientific contents. Using art as a successful means of presenting controversy in biotechnology has precedents in the *GenEthics* travelling exhibition by Marie Nordström in Sweden and in the *Biohistory Research Hall* by Keiko Nakamura in Osaka, Japan.

The design

In order to explain complicated principles and phenomena, it is necessary to make drastic simplifications, use elementary models and make basic comparisons. This may not always seem to be correct to a scientist. By simplifying complex issues we run the risk of distorting the 'scientific truth'. The fact that genes are invisible to the naked eye is a major challenge to the presentation of modern biotechnology. It is very difficult for lay people to understand something that they cannot see. Science-centre visitors have difficulty understanding that it is possible to base scientific conclusions on phenomena that are invisible to the naked eye.

Another problem is scale. For example, describing the size of a gene, the length of human DNA and how it fits into the cell and the amount of information that DNA contains is a challenge. We decided to use models, but I don't think they were very effective. Using computer animations might be a better way of explaining these issues, but we did not explore this route because we felt that we already had enough computers in the exhibition.

The design of an exhibit is extremely important, because the design itself sends many messages. If the design is unclear, then the message can be blurred or altered. On the other hand, the design is a powerful means of emphasising different issues. Because, in the public mind, genetic engineering is something that cannot be seen, cannot be touched and cannot quite be understood, it is important that the design of the exhibits connect as much as possible to everyday life and familiar objects. 'The Pigcow' exhibit, for example,

relates to a common product in Finland: a canned mixture of pork and beef. This is often served in garrisons to the young men doing compulsory military service. Another example of using familiar objects is the use of Helsinki telephone books in the exhibit on the Human Genome Project.

In the exhibit 'Let's Make a Super Baby', the problem was the lack of design. This exhibit consists of just two pairs of headphones and a black box with buttons to press. The original plan was to include two chairs shaped like a man and a woman. This was intended to emphasise the controversial nature of the ethical discussion. However, these chairs were never made.

Modern biotechnology is developing extremely rapidly. As a result, exhibits will soon become outdated. There are, however, ways to minimise this. One possibility is to present basic facts and principles, because they do not change as quickly. Another, often used method, is to take a retrospective view and to present milestones in the development of modern biotechnology. However, these two methods of interpretation do not provide an interesting enough exhibition. In the experience of Heureka, the most effective strategy is achieved by trying to look as far ahead as possible, presenting the leading edge of knowledge and addressing those universal issues that are relevant to the public.

Notes and references

1 von Troil, H, *Geeniteknisten Tuotteiden Hyväksyntä* (Helsinki: Biotekniikan ja Käymisteollisuuden Tutkimussäätiö, 1995), p94

2 Zechendorf, B, 'What the public thinks about biotechnology', *Bio/Technology*, 12 (1994), pp870–75

3 Marlier, E, INRA (Europe) European Coordination Office for the Commission of the European Community DGXII, *Eurobarometer 39.1, Biotechnology and Genetic Engineering: What Europeans Think about It in 1993* (Brussels: Commission of the European Community, 1993). The report is available from the Commission, Rue de la Loi 200, 1049 Brussels, Belgium.

4 Marlier, E

5 Persson, P-E, 'Science centres: dedicated to inquiry and exploration', *Physics World*, 9 (1996), pp55–56

6 Elomaa, P, Helariutta, Y, Griesbach, R J, Kotilainen, M, Seppänen, P and Teeri, T H, 'Transgene inactivation in

Petunia hybrida is influenced by the properties of the foreign gene', *Molecular and General Genetics*, 248 (1995), pp649–56

7 Zechendorf, B

8 Frewer, L and Shepherd, R, 'Ethical concerns and risk perceptions associated with different applications of genetic engineering: interrelationships with the perceived need for regulation of the technology', *Agriculture and Human Values*, 12 (1995), pp48–57

9 Marlier, E

10 von Troil, H

Creating *Gentechnik—Pro & Contra*: a balanced exhibition on genetic engineering

Mathis Brauchbar

Exhibitions on controversial issues such as genetic engineering call for an open-minded and balanced approach.

When genetic engineering became widely known outside the research laboratories in the 1980s, public discussion of risks and benefits of this new technology began. Whereas promoters of genetic engineering endeavour to describe the technique as a tool for the future solution of major problems such as disease or food shortages, the opponents of genetic engineering stress the risks and the yet unknown impact of a technology able to manipulate nature. As a result of clearly opposing arguments, public discussion on biotechnology and genetic engineering has become highly polarised.[1] This polarisation seems to be reflected in the media, which is the major source of public information on this technology. Thus, balanced examinations of the use of genetic engineering from case to case have rarely had any place in public discussions.[2]

This paper describes a novel approach to presenting controversial technology and science issues by means of an exhibition. The exhibition on biotechnology in Switzerland was initiated by our agency, Locher, Brauchbar & Partner AG, a communications and public-relations agency specialising in ecology, medicine and technology. As a mediator, Locher, Brauchbar & Partner aims to build bridges between different stakeholders, such as representatives of the administration, the economy and non-governmental organisations.

The object of the exhibition on biotechnology was to provide a credible source of information for interested lay people, in such a way that visitors would not feel that they were being persuaded to adopt a certain viewpoint. In addition, the exhibition aimed to show that,

although different parties sometimes had controversial views on the benefits and regulation of the applications of biotechnology, there were also many issues on which there was consensus. Therefore, the debate of experts on biotechnology was not always as polarised as it was portrayed to be in the media.

Despite vast media coverage of genetic engineering in Switzerland, the population's knowledge of the topic is still low.[3] But knowledge is a prerequisite for any individual to make informed decisions. 'Knowledge' does not imply that one has to know, for example, the structure of DNA. Judgments are based on values and confidence much more than on technological details. Therefore, communication on genetic engineering has mainly to deal with the fears of the public, which are generally of a moral or a political nature.[4]

Another crucial issue of communication is the credibility of the person or institution who provides the information.[5] Information, no matter how reliable, is useless if the addressed person does not trust the source of the information. Especially for complex topics, lay people (as well as experts) often have no means of challenging the accuracy of information.[6] Therefore, a decision-making process by lay people must be based on credible information. By implication, the source of information has to be transparent and sincere.

Initiating the concept

Our exhibition on genetic engineering endeavoured to inform visitors in a balanced and credible manner. The exhibition avoided blatantly aiming to enhance public acceptance of biotechnology, but offered a basis for visitors to make up their own minds. Consequently, there had to be a well-balanced advisory panel responsible for the exhibition, to guarantee credible information on the topic. This panel participated in the creation of the exhibition from the beginning: it was essential that panel members reflected on the form as well as on the content of the information to be presented. This meant that the members of the panel not only lent their names but also were responsible for the conception of the exhibition. In order to stress the commitment and thus the credibility of the panel, it was important that its members provided a major part of the financial support for the exhibition.

The initial stimulus was given by our agency to develop a long-term project to raise public awareness of genetic engineering. In the process of organising such an exhibition in Switzerland, opinion formers both for and against genetic engineering were invited to join the panel. At the end of this initiating process, five institutions constituted the panel. Represented by one delegate each were a group of pharmaceutical companies (Ciba, Roche and Sandoz), the largest Swiss consumer organisation (Konsumentinnen Forum), an environmental protection organisation (Swiss Society for Environmental Protection: the fourth largest in Switzerland with 22,000 members), and the Federal Office of Environment, Forests and Landscape. The canton (county) and town of Basel were represented by two delegates: the head of the Museum of Natural History and a representative from the department of education.[7]

In addition, the advisory panel included a mediator from Locher, Brauchbar & Partner, whose job it was to give input to the negotiating process, to propose solutions if problems arose and to propose the elements of the exhibition. The delegates on this initial panel, with the help of the mediator, negotiated the concept and the content of the exhibition. This panel received concrete proposals which each member discussed within their organisation. After this hearing process the proposals were discussed again and revised at a meeting of the panel. Only after a second hearing in the respective organisations were proposals approved.

Defining the content of the exhibition took about a year, with the panel meeting ten times. Besides the sometimes hard negotiating process, there was another factor that accounted for this rather slow way of proceeding: All members of the panel were representing organisations that had to be fully informed about every detail of the process and had to be convinced of the value of the exhibition. At the end of this year of hard work, a document was presented in which every aspect and every sentence of the exhibition was articulated and described in detail. Although the members of the panel had the opportunity of stepping down after the negotiating phase if they wanted to, they all elected to remain, and the construction of the exhibition began.

Elements of the exhibition

The exhibition, named *Gentechnik—Pro & Contra*, consisted of four parts. The first part dealt with heredity, genetics and the nature of genes. The second part demonstrated the laboratory work of genetic engineers. The third part showed ten applications of biotechnology already on the market, and ten applications which were in the stage of research or development. The last and most important part dealt with the continuing political debate on genetic engineering in Switzerland, addressing topics such as ethics, patenting, environmental risk, regulation, etc.

In order to make the exhibition attractive to visitors, the text was kept to a minimum. Instead, there were interactive games and video sequences that would attract and hold visitors' attention (figure 1). For further information, leaflets on specific topics and on general information, prepared by the members of the panel, were available.

The exhibition, which covered an area of 200–300 m², depending on the site, was designed to be suitable for touring. It was targeted at people interested in current

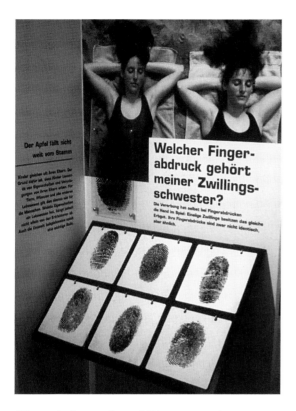

Figure 1. Interactive exhibits were an important part of the Gentechnik—Pro & Contra *exhibition*

political topics, scientific topics and/or genetic engineering. In addition, the exhibition was specifically designed to reach school groups, teaching professionals, students, civil servants/officials, and political and economic opinion formers.

In order to reach the target groups, the site for the exhibition was crucial. In Switzerland, museums of natural history are an obvious choice. There is one in nearly every city, they have a continuing tradition in educating science, they are institutionalised, and they are widely recognised by the public, especially among young families and school groups. There are very few science museums or science centres in Switzerland.

Locher, Brauchbar & Partner provided every museum that hosted the exhibition with support in the form of publicity, the organisation of guided tours for school groups

or other interested groups, and special events such as public debates. Usually, one to five public debates were arranged at which experts discussed specific topics on genetic engineering from a podium and with the public.

Touring

The exhibition *Gentechnik—Pro & Contra* opened in 1993 and had been shown in seven cities in Switzerland by the end of 1996.[8] Around 100,000 individuals and 600 school groups have visited the exhibition. Twenty public debates have been organised to date, with the level of attendance varying between 30 and 250 participants, depending on the topic and the location. The ages of the visitors were rather younger than the average population, because the exhibition especially addressed school classes and young families.

Evaluation and publicity

In order to give visitors the opportunity to express their opinion on the exhibition, they were invited to fill out a computer-based questionnaire. Around 15 per cent of the visitors took this opportunity and answered the questions. The following tentative conclusions can be drawn from the data, but it should be stressed that, as the sample was self-selecting, the results should be treated with caution.

The responses to the questionnaire suggest that visitors were neither clearly for nor decidedly against genetic engineering. Only one-third of the respondents had a definite opinion, while one-third was fairly in favour and one-third was fairly against genetic engineering. For people over the age of 65, however, this pattern did not hold. On average, males were more positively minded towards biotechnology than females. This is broadly consistent with general surveys in the European Union[9] and in Switzerland.[10] Visitors were asked to give their opinion on regulation in Switzerland. A fifth had no opinion, but half of those who did respond said that regulation was not strict enough—which is roughly the same proportion that considered genetic engineering to be dangerous.

About two-thirds of the visitors who completed the questionnaire considered the exhibition to be well balanced. This result shows that a majority of the visitors believed the exhibition to be credible and trustworthy. This conclusion was also reflected by the media response to the exhibition: not a single report questioned the objectivity of the panel or the balance of the content. The national and regional media acknowledged the exhibition's effort to present biotechnology in an unbiased way. Consequently, the media coverage focused more on the consensus among the parties rather than the differences.

Conclusions

Experience in Switzerland has shown that using a balanced advisory panel for an exhibition on a controversial technological

issue can result in a successful presentation of science. Essential to the exhibition's success were: the open-mindedness of panel members; the presence of a mediator whose competence in negotiating different standpoints was recognised by all the members of the panel; highly credible organisations and institutions represented on the panel; and sufficient time to go through the complex process of negotiating the exhibition content.

The scale of the success of the exhibition exceeded expectations, and it may lead to further collaboration in the long term. The members of the advisory panel agreed to consider future cooperation after the exhibition closed in December 1996.

In my opinion this strategy proved to be the right one for Switzerland, and it could be suitable for setting up exhibitions on controversial issues in other countries. However, the process must relate closely to the political circumstances in the country and to the prevailing state of public discussions on the issue.

Notes and references

1 Kliment, T, Renn, O and Hampel, J, 'Die Chancen und Risiken der Gentechnik aus der Sicht der Bevölkerung', in von Schell, Th and Moor, H (eds), *Biotechnologie—Gentechnik: Eine Chance für neue Industrien* (Berlin: Springer, 1995)
2 Stöcklin, S and Brauchbar, M, *Kommunikation über Bio- und Gentechnik bei Lebensmitteln in der Schweiz* (Bern: Schweizerischer Wissenschaftsrat, 1995)
3 Brauchbar, M, Locher, R and Wessels, H-P, *Wahrnehmung und Akzeptanz von Bio- und Gentechnologie bei Lebensmitteln* (Bern: Schweizerischer Wissenschaftsrat, 1995)
4 Ives, D, 'Public perception of biotechnology and novel foods: a review of survey evidence and the implications for risk communication', paper presented at the *Workshop on Novel Foods and Risk/Benefit Communication* (Schloss Reisensburg: Infratest Munich on behalf of the European Commission DG XII, 1995)
5 See, for example, Renn, O and Levine, D, *Credibility and Trust in Risk Communication* (Jüllich: Kernforschungszentrum, 1989).
6 Postman, N, *Technopoly* (New York: Alfred A Knopf, 1991)
7 The following institutions were members of the board: Schweizerische Gesellschaft für Umweltschutz; Konsumentinnenforum der Schweiz; Bundesamt für Umwelt, Wald und Landschaft; Erziehungsdepartement des Kantons Basel-Stadt; Interpharma (representing the companies Ciba, Roche and Sandoz).

8   The cities where the exhibition was shown were Basel, Chur, Zürich, Luzern, Aarau, St Gallen and Frauenfeld.

9   Marlier, E, 'Eurobarometer 35.1: opinions of Europeans on biotechnology in 1991', in Durant, J (ed), *Biotechnology in Public: A Review of Recent Research* (London: Science Museum, 1992); INRA (Europe), European Coordination Office, *Biotechnologie et Génie génétique: Ce qu'en pensent les Européens en 1993* (Bruxelles: Commission Européenne, 1993)

10 Brauchbar, M, Locher, R and Wessels, H-P

Gene Worlds: an international collaboration

Birte Hantke and Esther Schärer-Züblin

Gene Worlds *is a large project consisting of simultaneous exhibitions on contemporary biotechnology, organised across five museums.*

In 1998, five museums in Germany and Switzerland will unveil the results of a major international collaboration. *Gene Worlds* is a project comprising exhibitions examining genetic engineering and biotechnology from a variety of scientific and cultural perspectives. This paper describes the background to the collaboration, the common themes and the different approaches adopted by the partners.

The genetic engineering controversy in Germany

As progress has been made in molecular biology, genetics and biochemistry, new methods and laboratory techniques have been developed. One of these is genetic engineering. The term evokes mixed reactions among the general public. Advocates see it as a way of solving problems that so far have seemed insoluble, especially in the world of medicine, agriculture and environmental protection. It has also been seen as a means of ensuring that Germany remains an attractive place to do business. Its opponents, on the other hand, fear that its effects on individuals, the environment and society as a whole will be uncontrollable. In Germany in particular, voices warning that the Human Genome Project could become the basis for a 'new eugenics' are growing louder. 'Green' genetic engineering is barely accepted in Germany. The discussions about compulsory labelling of genetically altered foods reflect this attitude. Trial fields where transgenic plants are grown have to be guarded, and yet the fields are still often destroyed by militant opponents of genetic engineering. Since 1989, the industry

has increasingly moved its research and production centres abroad, where there have been few, if any, protests. The German government followed this development with concern and reacted accordingly. As a result, the strict German genetic engineering laws were relaxed in December 1993, and financial support for the Human Genome Project was considerably increased in 1996 in response to international pressure. In addition, more venture capital has been made available to encourage the setting up of small genetic engineering companies in Germany. At the same time, government information services are concentrating on increasing the level of acceptance among the public.

However, according to the well-known German genetic researcher Ernst-Ludwig Winnacker, the radical phase of the controversy over genetic engineering—the 'armoured car era'—in Germany is over. The public desperately need factual information and many scientists have now recognised the need for greater transparency in research and science. Scientists are more receptive to public-relations initiatives and the press than before.

The Swiss context

The people of Switzerland are bombarded on a daily basis by headlines on the subject of biotechnology, in particular genetic engineering. Genetically modified maize and soya beans are awaiting Swiss import approval for use in the food industry. Health-conscious citizens are at a loss about what to think. Objective information is often hard to come by and widespread emotional controversy does little to help people reach an informed opinion on the issue. Polls show that a large percentage of the population has, at best, only a superficial knowledge of the issues. A lack of adequate information means that people rely on preconceived notions and fears, and a general climate of technophobia has spread to include the use of genetic engineering. Public acceptance of genetic engineering in Switzerland varies according to the cultural region and correlates with different attitudes to nature that are rooted in history. In German-speaking areas, Germanic mind-sets prevail: the sacred forest of the Germanic people enjoins conservation of the given. This contrasts with the Latin world of the French-speaking part of Switzerland, which considers nature to be at its bidding.

A few years ago genetic engineering, which is now such a burning issue where food is concerned, hardly raised an eyebrow. There was certainly no talk of an 'initiative' on the subject, as is now in the offing. (The 'public initiative' and its better-known cousin the 'referendum' are the twin instruments of Swiss direct democracy. The difference is that the referendum allows citizens to reject measures approved by the legislature, while the initiative enables the people to introduce legislation directly.) In Switzerland, there was no hint of controversy or of an initiative when the idea of an exhibition on genetics and genetic engineering in biotechnology was first broached. The idea of an international exhibition project for this major theme made eminently good sense. It is in no way a child of the political situation in Switzerland, but its opening certainly has come at an awkward moment in the Swiss political scene.

In the US and in other parts of the world it may be hard to understand what the controversy is all about. In general, however, the issue can be seen as an example of scientific research that is the subject of heated discussion. The results of this discussion will have far-reaching implications not only for the world of science but also for society and the economy, as have subjects like ecology, energy technology and the neurological sciences.

Opportunities for museums

Authentic objects, sounds, smells, interactive models, experiments, artistic elements, audio-visual media, multimedia, games, written text and language: all of these are elements with which, ideally, modern exhibitions try to appeal to the visitor in every possible way. This is what makes exhibitions different from other educational events and often what stimulates the curiosity of visitors. In contrast to interest groups, whose approaches to the public are often clearly oriented towards increasing acceptance and sales, museums are regarded by the majority of visitors as independent. We would suggest that museums are therefore a particularly suitable platform for controversial debates and are also ideal for dealing with current research issues in public, and discussing them open-endedly.

In this context, the *Gene Worlds* cooperation scheme aims to handle the topic of genetic engineering in an interdisciplinary way in five special exhibitions in Germany and Switzerland in 1998. In addition, a number of institutions in Germany have taken the initiative in developing permanent exhibitions on genetic engineering: the Max Planck Institute for Breeding Research in Cologne (Cologne PUB), the Office for the Assessment of the Consequences of Engineering in the German Bundestag (Forum for Science and Technology, Göttingen), the Deutsches Museum, Munich (as a part of a new permanent exhibition), and the Max Delbrück Centre for Molecular Medicine, Berlin-Buch (*Transparent Laboratory*). The opening date for the *Transparent Laboratory* is expected to be 1998 and completion dates for the other projects lie even further in the future. Also, travelling exhibitions on genetic engineering are being planned for the Gene Centre in Munich and at the Ministry of Education, Science, Research and Technology.

The **Gene Worlds** *project and its inception*

The *Gene Worlds* project was initiated by the Alimentarium. Exploratory discussions were held with the Deutsches Hygiene-Museum (German Hygiene Museum) in Dresden. Personal contacts were of prime importance in finding other partners. The Kunst- und Ausstellungshalle (Art and Exhibition Hall) in Bonn and the Museum Mensch und Natur (Museum of Man and Nature) in Munich signed up. The Landesmuseum für Technik und Arbeit (Provincial Museum of Engineering and Work) in Mannheim, which had a project of its own in preparation on this theme, dropped plans for its own project in favour of taking part in the joint project. The project was eventually launched by six museums with a wide range of exhibition policies, and it was decided not to seek any more partners in order to avoid ponderous administrative problems. Sadly, a Dutch museum—Museum Boerhaave—later had to pull out of the project because planning for a 1997 exhibition on the same theme was already under way in Leiden.

Since the middle of 1996, each museum has had its own project leader. The Kunst- und Ausstellungshalle set up an advisory panel consisting of a moral philosopher, a social ethicist, a biochemist and a human geneticist. They drew up a catalogue of keywords touching on all the major topics involved. It is important to note, however, that there is no one consensual concept that is duplicated or travels around as a touring exhibition. What the project members have agreed on is a common project theme with a variety of sub-themes, developed wherever possible in association with each other, each of which is integrated with the individual museum's exhibition. This common theme is generally illustrated in a highly critical fashion, but with a positive undertone. What is pioneering, and also distinctly risky, about the project is the fact that five partners, with totally different expectations of an exhibition and varied visitor profiles, are working on a project that is attuned to the particular concerns of the individual museums while sharing the same publicity campaigns. This arrangement allows for cooperation, but engenders competition, for example in attempts to obtain original objects for exhibitions, such as Gregor Mendel's microscope, his eyeglasses, or the original Watson–Crick model of the double helix. Joint activities are planned by the various institutions, such as press and publicity, publications and educational ventures. For example, a joint catalogue with articles on various aspects of the main themes is currently being prepared. Plans are also being made for joint events at which well-known anthropologists, art scholars, natural scientists and representatives of industry and politics will be talking about issues relating to genetic engineering and discussing them with the public.

On 25 March 1998, a major joint press conference will be held in Bonn. The various exhibitions will be opened the following day in Bonn, Dresden, Mannheim, Munich and Vevey. We are all keen to see how people will take to this new form of cooperation between museums.

Individual participants in **Gene Worlds**

As already mentioned, each of the five institutions in the *Gene Worlds* project is creating its own *Gene Worlds* exhibition (table 1). The resultant varying exhibitions have different thematic focal points and intentionally use educational methods that often vary considerably. All the institutions see themselves as mediators between scientists and lay people and as a forum for debating questions of relevance to society.

Alimentarium

The Alimentarium, a food museum in Vevey, Switzerland, which opened in 1985, is a Nestlé foundation. It depicts scientific, ethnological and historical aspects of food in an interdisciplinary context (figure 1). The museum puts on two to three temporary

Table 1. Collaborators in the *Gene Worlds* project

Museum and subject area	Annual visitor numbers and visitor profile	Exhibition title	Main themes	Area (m^2)
Alimentarium Nutrition: nature and culture	50,000, 1:1 ratio of adults to children	*Au fil du gènes*	Epistemology of breeding and genetic engineering, molecular biology, applications in agriculture and food, consumers' fears and visions	450
Deutsches Hygiene-Museum Humankind: body and culture	200,000 exhibition visitors, 100,000 visitors to events, 2:3 ratio of adults to children	*Workshop Man?*	History of and methods in genetics and genetic engineering, health, applications of genetic engineering to medicine, society and ethics	1000
Kunst- und Ausstellungshalle Art, natural and cultural sciences	600,000*	Under discussion	Structure vs information, nature vs creation, evolution, health, environment, nutrition, market forces, visions, temptations	1600
Landesmuseum für Technik und Arbeit History of engineering, social history	230,000, 1:1 ratio of adults to children	*Life from the Laboratory*	History of and methods in genetics and genetic engineering, working with genes, selected genetic engineering applications, opportunities and risks	800
Museum Mensch und Natur Humankind, nature, evolution and ecology	280,000, 1:1 ratio of adults to children	*Reaching for the ABC of Life*	History of and methods in genetics and genetic engineering, selected genetic engineering applications	600

*Visitor profile not known

exhibitions a year, in addition to its permanent exhibition.

In keeping with its tradition, the Alimentarium wants the *Gene Worlds* exhibition to appeal to all types of visitors, from those with a high degree of acceptance, through those showing critical detachment, to those advocating total rejection of applied genetic engineering in the area of food and nutrition. The museum hopes it will spur people to think about the issue for themselves.

Deutsches Hygiene-Museum (German Hygiene Museum)

Founded in 1912 by Odol manufacturer Karl August Lingner, the Deutsches Hygiene-Museum in Dresden is dedicated to

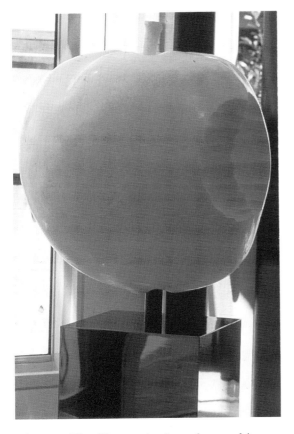

Figure 1. The Alimentarium's apple carved in white marble symbolises the theme of 'man and food'

humankind, the human body, health, society and the environment. It uses its major special exhibitions as a forum for discussion on current social issues.

The aim of the exhibition *Gene Worlds: Workshop Man?* is to illustrate the controversial relationship between illness and health, and to explain the basic principles and methods of genetics. The exhibition will cover possible applications of genetic engineering to medicine, and present examples of particular genetically engineered products and promising strategies for tackling previously incurable illnesses. In addition, social issues, such as different ethical viewpoints, possible social implications, questions of patent law and the economy will be handled in an objective way. Interactive demonstrations, such as a large-scale interactive demonstration of protein biosynthesis, working models, molecular biology equipment animated using audiovisual media, stimulating artistic installations and interviews will provide visitors to the Deutsches Hygiene-Museum with the opportunity to gather information in an entertaining way. A number of discussion rooms will be integrated into the exhibition, and visitors' books will offer people an opportunity to tell other visitors what they think or to look up other people's views. An Internet site in the exhibition will open up access to discussion forums throughout the world.

Kunst- und Ausstellungshalle der Bundesrepublik Deutschland (Art and Exhibition Centre of the Federal Republic of Germany)

The programme of the Kunst- und Ausstellungshalle, which opened in 1992, covers a broad range of subjects in which natural science exhibitions play a part along with art and cultural history. The Kunst- und Ausstellungshalle will be inviting visitors to learn and reflect on science, its technologies and its areas of applications. The findings of genetics and their importance for the reality and the imagination of human beings will be presented from the viewpoint of cultural history. The concept of this exhibition focuses less on simply displaying techniques and scientific results. Its objective is to present dates, facts and images in a thematic framework that is relevant to the general public. The choice of topics (from the fields of nutrition, agriculture and medicine, for example) is problem oriented. Special features include forms of presentation based on contemporary thematically related works of art.

Landesmuseum für Technik und Arbeit (Regional Museum of Engineering and Production)

Since its opening in 1990, the Landesmuseum für Technik und Arbeit in Mannheim has emphasised the presentation of industry and technology from the perspective of social history and the study of civilisation. The exhibition *Life from the Laboratory* will develop this theme and show that, with genetic engineering, the ability to change and perfect life has reached a new level.

Historical objects, documents or scenes will be on view in display cabinets. The main part of the exhibition will cover the subject of working with genes, where visitors can watch demonstrations of some of the most important genetic engineering processes and try them for themselves. These demonstrations will form part of life-sized displays in an areas on work in the laboratory, health and illness, environment and safety, society and politics. In the open areas around the Landesmuseum visitors can encounter the world of agriculture and plant genetics. The third part of the exhibition will look at the prospects for genetic engineering and visions for the future. The 'Visions Market' will pose the question: is the whole world turning into a laboratory?

Museum Mensch und Natur (Museum of Humankind and Nature)

The Museum Mensch und Natur in Munich was opened in 1990. Its topics are the history of the Earth and life, the diversity of organisms, and man as a part of nature as well

as a designer of it. The aim is to impart knowledge on natural history in a didactic as well as an entertaining manner for different ages and levels of education.

The exhibition at the Museum Mensch und Natur is entitled *Reaching for the ABC of Life*. Tracing the central theme of the story of the discovery of DNA should lead to insights into genetics and cellular biology: from the general hereditary developments to the molecular mechanisms of the control of the processes of life. The current level and prospects for applications will be presented on the basis of selected models for medicine, food production and preserving biodiversity. Models and explanatory videos that bring the topics to life, interactive quizzes and games, fascinating facts about the 'pets of the genetic world', and selected examples of applications will all be aimed at giving the visitor an idea of genetics and the possibilities of genetic engineering.

Interpretation

Modern branches of natural science, such as molecular genetics, frequently use abstract concepts. The objects that are being studied are usually so small that they can be measured or depicted only indirectly. Much of the special equipment used is highly complex and the results need to be interpreted by experienced scientists. Such specialists think in models and use formulae and abbreviations much of the time. These factors all represent a huge challenge to curators of exhibitions. The use of original photographic and film material and of interactive models was the subject of much discussion among the museums cooperating in the *Gene Worlds* project. Opinions differed primarily about how to define the extent to which things should be simplified, about the extent to which explanations should be given, and how best to put across the 'fascination of research'. There was a lively exchange of views and it was agreed that the individual museums should continue to pursue their different ways of communicating knowledge in line with their different aims and audiences. This is largely what makes this cooperative venture so special and so attractive.

The example of the 'cell' illustrates in particular how differently scientific photographs and film, three-dimensional models and explanations are handled within our project. The Deutsches Hygiene-Museum owns a large glass model of a human cell made in 1987, which is 2×1 m in size (figure 2). The individual organelles are selected by pushing a button; they light up in the model and a sound recording tells the visitor about their function. The Deutsches Hygiene-Museum will be displaying this impressive exhibit in its *Gene Worlds* exhibition. It conveys an idea of

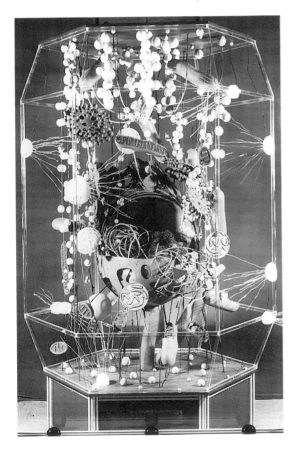

Figure 2. The Deutsches Hygiene-Museum's model of a human cell will be on display in their Gene Worlds *exhibition*

the structure of a cell and the relationships between the sizes of the cell organelles and structures. Although it is a static model, it provides lay people with a means of imagining and understanding processes within cells. Films and slide preparations are shown around the large model to illustrate the processes in living cells and cell dynamics. The Museum Mensch und Natur will also be making use of a cell model. Like other models, it will be explained in videos by scientists and brought to life through films shot through a microscope.

The Kunst- und Ausstellungshalle has chosen not to show any models of cells; instead, it wants to use the latest pictures taken from research for its teaching purposes. The aim is to encourage visitors to think about the issues behind the choice of the research method and method of depiction. The Alimentarium intends to combine the use of models with the latest scientific video material. This approach enables visitors to understand the evolution of cell models as well as the complexity of cell dynamics, without losing their sense of wonder.

The Landesmuseum für Technik und Arbeit is not planning to include any major exhibit on the cell. The complexity and size of the cell will be conveyed in the context of an historic scientific discovery, using original material and presentations.

The different presentations of protein biosynthesis show the two different attitudes to three-dimensional interactive models within *Gene Worlds*. The Deutsches Hygiene-Museum is planning a 4×6 m presentation of the cell to help visitors to understand the principle of protein biosynthesis and thus the principle of translation of hereditary information into proteins. The visitor takes on the role of enzymes and other supportive factors, and can make a messenger copy of the hereditary information in the cell nucleus using special three-dimensional modules. This can be transported through a pore in the nuclear membrane into the cell plasma. Using loaded carrier components, visitors can then gradually build a protein or a polypeptide chain. When the activity has been successfully completed, a short film shows what function the polypeptide has in the body. The fact that protein biosynthesis in the human cell is more complex than appears in the model and is subject to molecular control is explained in additional video animations and drawings.

Concerns have been expressed about whether models of invisible and intangible structures give an accurate picture of the possibilities of intervening in living systems.[1] By allowing visitors to manipulate models, a misleading impression may be given about the abilities of scientists to 'design' living things. To overcome this objection, the Kunst- und Ausstellungshalle wishes to use video animations where the object size makes it difficult to display, and where functional interconnections and processes are to be conveyed. The other museums will be presenting protein biosynthesis in the form of computer animations, mainly for reasons of cost.

The exhibitions will be financed by the museums in different ways: the Alimentarium is a Nestlé foundation, and the Kunst- und Ausstellungshalle receives government funding. The other three museums are negotiating with various sponsors.

Conclusion

There are advantages to an international collaboration of this type from both an organisational and an interpretative point of view. By exchanging information and exhibition material, and creating a common publication, costs are reduced. Furthermore, a collective approach is helpful for interpretating controversial issues like genetic engineering. This does not necessarily mean a reduction in diversity, however. Within the *Gene Worlds* project, genetic engineering is seen from a variety of points of view. Different exhibitions at five museums are able to show many more aspects of genetic engineering than a single museum could ever hope to cover. The uniqueness of this event, and the fact that all five exhibitions will open on the same day, makes it likely that it will draw the attention of local, national and even international audiences and media.

Notes and references

1 Kridlo, S, Collaborator on the concept for Kunst- und Ausstellungshalle, personal correspondence (November 1996)

Diving into the Gene Pool, an exhibition at the Exploratorium

Charles Carlson

This exhibition on the Human Genome Project and genetics serves as a case study for the presentation of interactive biological exhibits.

For the past 25 years, the Exploratorium has sought to develop interactive biology-based science exhibits, exhibitions and displays. During this period we have developed more than 80 different exhibits, focused on many different aspects of biology, from molecules to organisms. Recently, we developed an exhibition on genetic science, the Human Genome Project (HGP), and its associated ethical, legal and social impacts, called *Diving into the Gene Pool*.

The 28 individual exhibits in *Diving into the Gene Pool* took approximately four years to develop at a cost of about £400,000.[1] Starting in April 1995, the 325 m[2] exhibition ran for five months and attracted approximately 178,000 visitors. Fifty-eight thousand of these visitors came specifically to see the exhibition.[2] To complement the exhibition, we conducted a three-part bioethics lecture series, and organised five family-oriented weekends which involved programmed activities and mini-lectures. Today, many of the exhibits from *Diving into the Gene Pool* remain on display, forming an attractive subset on genetics in the Exploratorium's more permanent biology exhibit area. We are currently in the process of creating a new presentation of biology at the Exploratorium. The experience gained from *Diving into the Gene Pool* will help to inform the reassessment and revision of how we present biology in general—and genetics in particular—as we plan future initiatives.

Background

In keeping with the Exploratorium's overall philosophy, our primary emphasis in creating *Diving into the Gene Pool* was to give visitors access to demonstrable phenomena in a learning environment. Creating the exhibits proved to be a rich and challenging endeavour, from both intellectual and technical perspectives. I believe the exhibition can serve as an exemplar for the exploration of many of the issues connected with displaying the biological sciences and biotechnology.

The HGP is a large-scale, worldwide scientific effort aimed at sequencing all of the genes in the entire human genetic code of some 3 billion nucleotide bases by the year 2005.[3] The understanding that comes from sequencing this DNA code has radically increased our understanding of human life, and provided science and medicine with an unprecedented set of tools by which we might understand other life forms, create new life forms, alter existing forms, and both diagnose and cure disease. Because of the potentially powerful and pervasive nature of the information gained from this and other related projects, numerous ethical, legal and social issues arise for presentation, discussion and decision making in the public forum. By its nature, the HGP creates a profound societal need for informed decision making, but the abstract nature of its science obscures its presentation.

The science and technology which underlies genetics and the HGP rests upon our understanding of deoxyribonucleic acid (DNA) and its associated molecular pathways. These microscopic and elegantly simple biochemical relationships create a common thread between organisms. Much of this same molecular science finds practical application in biotechnology, giving rise to new and powerful tools and methodologies in the health-care and pharmaceutical industries. This ongoing discovery, mapping and exploitation of nature creates a common set of problems associated with the presentation of science at the nexus of theoretical and applied knowledge. A discussion of genetics and the HGP has relevance for much of biology, biotechnology, chemistry, and perhaps areas of physics such as nuclear physics and quantum mechanics, which also deal largely with the sub-microscopic.

Issues in presenting molecular biology

From an interpretive perspective, genetics and its related biology bestow a series of barriers to informal presentation. While we experience the manifestations of DNA's codes as various life forms from micro- to macroscopic, its generalised molecular biology resists all forms of direct visitor-friendly observation, manipulation and experience. DNA is too small to see with the unaided eye, it presents a superficially simple yet complex code of information and effects of this information may not always be obvious, or expressed within a practically demonstrable time frame. For example, a tiny test tube might hold the genetic codes for 1000 dogs in a drop of clear liquid. But a museum visitor looking at the same test tube does not experience 1000 dogs, DNA or anything particularly interesting, exciting or compelling. The visitor must take on faith that the tiny clear drop of liquid in a tiny tube holds the template for 1000 dogs. A bacterium producing a 'wonder drug' does so in relative secrecy, seemingly lost in a bubbling soup of fermenting growth media. A mutation may not always be obvious in the phenotype, and the chemistry of our own existence frequently defies a satisfying rational explanation.

By its nature, most of the fundamental aspects of life reside beyond the range of our directly experienced world, and serve as a source of fascination and wonder. Yet because of the complex and intangible nature of this material, the number of potentially productive experiences by museum visitors has limits. In a museum, where a somewhat non-sequenced, randomised order of experiences typify an average visit, the presentation of a more linearly organised set of concepts and facts becomes dauntingly difficult. Visitors find it hard to make meaning without some appropriate previous experiences or frame of reference.

Therefore, any exhibit about DNA and the secrets of life necessarily involves an array of scientific tools, structural models and indirect observation techniques coupled with logical inference. Understanding biology demands a careful selection and a somewhat sequential presentation of knowledge. Making a meaningful visitor experience may require the same approach, or a much higher degree of mediation through a demonstrator. This demand runs counter to the typical familiar physical science and psychophysics presentations found at most science centres.[4]

At the other end of the spectrum, in sharp contrast to the unified abstract molecular details of the DNA dogma and molecular biology, there is an almost overly rich world of experiential complexity at the level of organisms. People tend to be endlessly fascinated by all forms of life and its myriad variations. Each form contains an entire somewhat specific history; a world in itself. At this level, visitors experience the organism and not changes in DNA code or variations in proteins.[4] Generally, the larger and more complex the association of a cell, the less predictable and repeatable its behaviour and the higher its demand for resources. Extracting a highly reliable, focused demonstration or display from an organism, or even cells, can prove more than a trivial problem. To twist an old saying, one cannot see the organisms for the DNA and one cannot see the DNA for the organisms. Predictable general concepts are difficult to display successfully at the level of a complex system which tends to be unpredictable.

In addition to problems related specifically to the nature of genetics, a more general series of issues must be dealt with in biology and chemistry. These include consumables and additional maintenance requirements integral to any display using living organisms or chemical reactions. A simple chemical reaction requires a supply of reactants and generates a product which usually cannot be recycled and must be properly disposed of. Living biological tissue preparations require refreshing at intervals to function properly. To attain a level of functionality, most chemical and biological exhibit preparations require a dedicated and highly trained support staff to carry out the development, maintenance and service requirements. In addition to these practical aspects, ethical issues and considerations are part of any display using living tissues and organisms. For example, a display that uses people and their genetic characteristics or genes in a demonstration of DNA fingerprinting bypasses issues related to tissue refreshment and sample availability, but will almost always involve other practical, safety and ethical considerations.

Because interactive presentations on biology in general—and genetics and biotechnology in particular—have many inherent problems, serious consideration must be given to the rationale for presenting these subjects. This rationale can be delineated on both a theoretical and a practical level. Fundamentally, molecular biology and genetics provide keys to understanding life. Because people have a fascination with the organisms and processes of life and its origins, they tend to find genetics and molecular biology intrinsically interesting. These are traditional subjects in school and knowledge of them is part of a well-rounded education. In addition, genetics and molecular biology may be directly relevant to visitors' experiences, by providing an explanation for diseases and heritable genetic conditions, so there is direct medical relevance.

Pragmatically, molecular biology, genetics, immunology, microbiology and related sciences will provide the careers of the future as a nascent biotechnology industry grows to fruition. Government and private support for the HGP provide further impetus for the display of this science and technology. Biological engineering, the rapid rate of gene-related discoveries, and the creation of transgenic organisms are featured almost daily in the popular press. A working knowledge of the basics of biology is critical to understanding such announcements as the discovery of a gene that controls sexual orientation, the amount of body fat, predisposition for breast cancer, the insertion of a gene in a soya plant or the creation of transgenic organisms.

Because of its direct relevance to all life, it is important that the public are informed about genetics and biotechnology on a variety of levels. Many of the decisions that people will be faced with will require informed opinion, for which they need to be exposed to the science and relevant social, political and ethical issues that are part of the fabric of culture. Because museums have a role in providing both documentation and informal accessible learning, they may be useful in broad-based public education.

Diving into the Gene Pool

In creating *Diving into the Gene Pool*, our approach to presenting genetic science and the HGP focused on providing a strong scientific backdrop; picking out those areas we felt were essential for understanding of the scientific background; exploring the progress and purpose of the HGP; and providing a visitor framework for discussion and reflection on the associated ethical, legal and social issues. We divided the exhibition into five thematic areas: **1** DNA, the molecule of genetic information, **2** tools and the technology of genetic science, **3** lessons from other species, **4** families, sex, and human inheritance, **5** points of view/ethics resource area. Each area was denoted by a colourful yellow triangular column standing 2.4 m tall, that gave the area title and showed a large graphic presentation which described relevant aspects of genetics related to that exhibition area.

The thematic groupings provided an overall organisational framework that spanned a wide

Figure 1. The introductory panel was designed to convey the notion that the theme of the exhibition was relevant to everyone

range of genetic science, from molecular biology to applied genetics to related ethical issues. The core educational messages of the exhibits included:

- Every living thing is a product of a combination of inherited characteristics, passed from one generation to the next, and environmental influences (figure 1).
- The sequence of bases in DNA molecules contains the complete instructional blueprints for making an individual; these blueprints are contained in their entirety within almost every cell of every organism.
- Most multicellular organisms have unique genomes, yet all organisms have DNA as their genetic information and share some common DNA sequences.
- DNA's instructions are organised into functional subgroups called genes, which carry the instructions necessary for encoding proteins. Many traits—intelligence, hair colour, susceptibility to heart disease or breast cancer—are produced by a combination of genes and environment.
- The process of mutation—spontaneous random changes in DNA sequences—can cause changes in an organism's physical characteristics and/or in the physical characteristics of its progeny.
- There are many processes by which genetic variations enter and remain in gene pools; these include spontaneous mutation and natural selection, selective breeding, and recombinant technologies such as gene therapy.

In the first area, entitled 'DNA, the Molecule of Genetic Information', for example, we devised a graphic that compared a living cell to the planet Earth, chromosomes to countries, genes to cities and individual nucleotide bases to street addresses. We included free-standing exhibits such as: a 1 billion scale model of a short section of DNA which extended from the floor 6.7 m into the air; a visitor-operated computer-based molecular modelling station, 'Molecular Library', displaying DNA, nucleotide bases and various proteins; and a large, simplified, mechanical model of DNA demonstrating the steps required for the expression into a segment of a protein, called 'Protein Production Line'. Through these exhibits we hoped to cover most of the basic nature of the DNA molecule, from its location within a cell to its size, structure and function. Similarly, within all the other areas we attempted to provide a mix of both graphics and exhibit activities that best illuminated the five themes.

Graphics and exhibits were created for all the other areas with an emphasis on actual research materials and organisms. Some of the highlights are outlined below.

- Dancing DNA. Strands of viral DNA, labelled with a fluorescing compound, were placed in a miniature agarose gel under a fluorescent microscope, so that strands of DNA might be observed through a video camera and monitored at a magnification of × 5000. We then added a variable electrical field to this preparation so that these strands of DNA could be manipulated by visitors. Visitors could move the DNA through the gel by changing the electrical field via a control stick.

- DNA Fingerprinting. Complementing and adjacent to this microscopic view of DNA, we ran a fully functional laboratory electro-phoretic gel used for separating DNA fragments by size in a simulated DNA fingerprint preparation. Visitors were asked to match the various samples to determine which DNA samples were identical, in much the same way as a forensic scientist would do.
- Mutant Fruit Flies. We exhibited living wild-type and mutant fruit flies at several exhibit stations, to demonstrate the genetic variations across a species, and highlight the role that these organisms have in research.
- Blood Typing. This exhibit utilised human blood and monoclonal antibodies to demon-strate both the process of blood typing and the relationships between genotype and phenotype in humans. We also interactively displayed the morphological changes that happen in response to changing oxygen concentrations in human blood cells in people with sickle-cell disease.

A number of computer-based exhibits functioned either as stand-alone exhibits or adjunctively as supporting interactive graphics. One of these, 'Musical Mutants', displayed the matched DNA and amino-acid sequences (amino acids being the subunits that compose a protein) of a homologous protein found in four different organisms. The general idea was to illustrate the concept that variations among organisms rest on differences in their DNA and corresponding protein sequences, and to do this in a fun and interesting way. To accomplish this, we assigned each amino acid a particular note and gave visitors the ability to assign the sounds of particular musical instruments. To operate the exhibit, visitors played through the amino-acid sequence via a control on the computer screen, listening to the amino-acid sequence 'music' on one of four audio speakers, each of which corresponded to the protein of an organism. The music sounded melodious when the sequences between organisms matched and discordant when the sequences varied.

To demonstrate the uniqueness of every individual, we created a ten-question genetic quiz, called 'Genetic Inheritance'. As an individual worked through a series of yes-or-no genetic trait questions, such as whether their ear lobe is attached to the base of the ear or forms a distinct lobe, if they have blue or brown eyes, etc., the computer tabulated the results so that visitors could compare their genetic profile with others who had entered data. Even though the traits were chosen for having an approximately equal distribution in the population, in just ten questions visitors found that they were part of a similar cohort of only 0.2 to 0.3 per cent of the total number of visitors that had responded.

Infused throughout the exhibition, we made reference to a variety of ethical, social and political issues which might arise from the HGP. Using colourful graphics, a large-scale human gene map of 100 genes and their locations on chromosomes, a visitor-activated graphic called 'Mutant Detector' (which proved extremely popular, figure 2) and beautiful photographs of chemically painted chromo-somes (fluorescently labelled in-situ hybrid-isations, or FISH), we drew attention to human variation and various disease condi-tions.

In the 'Points of View/Ethics' resource area, we provided binders containing several hundred recent newspaper and magazine articles on genetic discoveries and ethical, legal and social issues; an Internet forum for a moderated, threaded discussion; and four talk-back boards that focused on particular bio-ethical dilemmas to which we invited visitors to make written responses. The dilemmas included presymptomatic testing for the gene for Huntington's chorea, the labelling of genetically engineered food and boundaries of germ-line therapy. The visitor feedback exhibits proved to be surprisingly popular.

Exhibition evaluation

The evaluation of *Diving into the Gene Pool* was based on visitor interviews, and staff

Figure 2. By walking through an activated gate, visitors discover that everyone carries mutations in their own DNA

and evaluator observations of visitor interaction with exhibits. We did not conduct a truly comprehensive evaluation of the exhibition because of financial and other resource constraints, so the evaluation findings and results need to be interpreted with a degree of circumspection.

During its five-month run, *Diving into the Gene Pool* generated widespread public and media interest, with coverage by more than 30 print and broadcast media, including local media as well as CNN and National Public Radio. An estimated 178,000 people (65 per cent of 274,000 total museum visitors) attended the exhibition, and 57,500 (21 per cent) stated that *Diving into the Gene Pool* was the sole reason for their visit. Of the visitors

who attended the exhibition, 80 per cent rated the exhibition as either 'excellent' (24 per cent) or 'good' (56 per cent). Approximately 28,000 children on school field trips visited the exhibition. Visits to the five thematic areas focusing on science tended to be relatively evenly distributed.[6]

The evaluation indicated that visitors found the individual exhibits fascinating and informative. Somewhat surprisingly, the results also indicated that many visitors intuitively understood the connections between the various thematic sections, although few were able explicitly to verbalise these connections, which we had intended to convey through the exhibition graphics. These results highlight the difficult nature of presenting complex and abstract subject material in the context of an unstructured museum visit. In sharp contrast, a guided or facilitated tour dramatically increased the visitor's overall understanding of various thematic areas. On average, visitors spent about eight minutes in the entire exhibition, with two-thirds only briefly stopping to look at one or two exhibits; the remaining visitors spent a great deal more time, some up to 32 minutes. The average visit time increased dramatically to slightly over an hour when facilitated by a tour.

While such a small amount of time generally spent visiting *Diving into the Gene Pool* may seem dismal on first examination, a comparison with a visit to the entire Exploratorium provides for a more meaningful comparison. The Exploratorium contains about 574 exhibits. Most visitors spend about two hours visiting these exhibits at a rate of about 13 seconds per exhibit. Visitors at *Diving into the Gene Pool* spent about 17 seconds per exhibit visiting 28 exhibits. This suggests that there is probably not a significant difference in usage patterns between the exhibition areas. Obviously, visitors do not visit every exhibit, but are picking and choosing in an area, depending upon their tastes and predilections. This observation correlates well with studies of visitor behaviour at other museums.[7]

Another observation made during the development of the exhibit 'Protein Production Line' was the correlation between educational background and exhibit appeal and usage. Visitors that used the exhibit as intended by the exhibit developers and found it appealing, educational and instructive, tended to be educated to beyond high-school-level biology. This correlation between educational background and exhibit preferences undoubtedly characterises much of a visitor's exhibit usage.

Another informal observation made about individual exhibits is that visitors found exhibits with living creatures both interesting and accessible. At the 'Find the Fish' zebrafish exhibit, visitors of all ages greatly enjoyed finding mutant fish among the wild-types. Likewise, the living mutant fruit flies continue to attract and fascinate visitors.

An unanticipated result of the exhibition was the large quantity and variety of public comments recorded as part of the 'Points of View' exhibit. This section of the exhibition contained four hypothetical stories presenting visitors with ethical choices about genetic research, engineering, notification and counselling. Visitors were asked to write their responses on cards compiled during the exhibition. 'Points of View' generated more than 5000 written responses, providing staff with data reflecting visitors' views on genetics issues. Sixty-two per cent of these responses were narrative remarks that took some time to compose. The magnitude and quality of this response indicates the tremendous interest among the Exploratorium's audience in the topic of applied genetic science and associated ethical issues. This high level of public interest in issues raised by the HGP may well have prompted people to become better informed about the science of genetics.

Conclusions

Diving into the Gene Pool proved to be a challenging exhibition to develop because of the abstract nature and quality of the subject material, and as such represents a prime example of the presentation of current areas of scientific research and development. In our approach to the topic, we attempted to blend interesting scientific examples and some art with a series of important educational messages. We achieved some measure of success in this presentation. Although the exhibition did not prove to be powerfully attractive from an audience-development perspective, most visitors had a positive experience. Somewhat surprisingly, the ethical exhibits in the exhibition proved to be extremely popular with a wide segment of the audience. In the evaluation we demonstrated that exhibits, or collections of exhibits, do not seem to present a good venue for the demonstration of abstract concepts or metathemes because visitors tend to experience a more compartmentalised series of phenomena. This remains a difficult area for informal learning environments. Our experience with *Diving into the Gene Pool* provides an array of relevant material for the development of a future exhibition on genetics and the HGP, and will prove useful in the development of future informal learning environments.

Notes and references

1   *Diving into the Gene Pool* was supported by the United States Department of Energy, and Genentech Foundation for Biomedical Sciences.
2   These unpublished data are based on an evaluation conducted by Inverness Research Associates in June 1995, and a visitor evaluation survey conducted by Morey and Associates, 1995.
3   *Human Genome News,* 7 (September–December 1995); US Department of Energy, *Human Genome Report* (1991–92); http://www.ornl.gov/TechResources/ Human_Genome

4 Visitor evaluation surveys by Inverness Research Associates (1995)

5 Anderson, P and Cook Roe, B, 'Museum impact and evaluation study', *Roles of Affect in the Museum Visit and Ways of Assessing Them* (Chicago: Museum of Science and Industry, 1993); Serrell, B, Beverly Serrell Associates, Chicago, personal communication (1997)

6 Visitor evaluation surveys by Inverness Research Associates (1995); Morey and Associates (1995).

7 Dierking, L and Falk, J, 'Family behaviour and learning in informal science settings: a review of the research', *Science Education,* 78 (1994), pp57–72; Falk, J, 'Analysis of the behaviour of family visitors in natural history museums', *Curator,* 34 (1991), pp44–50

Rapporteur's remarks

Contemporary science in museums and science centres: concluding remarks

Per-Edvin Persson

What are the conclusions? The rapporteur of the Here and Now *conference summarises the key points that emerged from the meeting.*

The papers of this conference, and the ensuing discussions, have painted a detailed picture of the various aspects of presenting modern or contemporary science to the general public. The two key questions with which this conference has dealt are: what special problems apply to the contemporaneity issue? And what is special for science museums and science centres in this context? First, we have to look at the whole picture, as Simon Schaffer did in his keynote address. We have to position science, science museums and science centres in society and history—that is, put science into historical perspective. Joost Douma of newMetropolis, Amsterdam, said in his talk that science museums and science centres are children of their time.[1] This is, of course, true; but it is equally true that science itself is a child of its time.[2]

As we are trying to paint the whole picture, it is important to address the question of why we need 'public understanding of science'. I think the reasons are well known and well established, although this has not been a particular focus of this conference. These issues are inherent in the themes that we have been discussing, and I recently dealt with them elsewhere.[3] We need public understanding of science because there is a need for an enlightened debate and decision making in society, especially related to science and technology. We think it is fair to provide information to taxpayers about how their money is being spent, and it is in the interests of consumers if they make well-informed choices, so why should they not be told how research funds are being spent? And then, of course, it is good for the scientists themselves to encourage public understanding initiatives, as it tends to help them to solicit funding for research.

So we need public understanding of science, but why do we need science museums and science centres? What added value do they bring in this context? Science museums and science centres involve visitors in real experiments, and they are the only medium that can do this (except, perhaps, a visit to a research institute during an open day). Studies show that science museums and science centres are worthwhile, and a visit is seen as a positive social event.[4] Research also shows that exhibitions enhance the motivation of school children.[5] Then, there is the rather compelling evidence in rising numbers of visitors.

There is a more elementary way of explaining why we need science museums and science centres. Good decisions require some sort of wisdom. The problem in the modern world is that there is abundant information available but, unfortunately, rather little wisdom around. Wisdom requires structured knowledge, which usually involves study. In order to study, you need to do some work. (All of this is, of course, evident from the second law of thermodynamics.) When work is performed by human beings it requires motivation. Science museums and science centres can provide this motivation. This really positions science museums and science centres in the fields of education and the public understanding of science. We are not really in the information business: we are in the motivation business.

Considering the big picture for science museums and science centres in this regard, we can ask what do science museums and science centres do best? The short answer is 'exhibitions'. Of course, there are educational programmes, public programmes, marketing, media relations and so on, but on the whole the interactive exhibition is the trademark of a science centre and a science museum. An exhibition is really a medium. Consequently, I was not at all surprised when Jana Bennett presented her interesting table on the criteria that apply to the success or failure of television programmes (see p51). They apply equally well to our exhibitions. I do not think that she meant that those things which are likely to fail should be omitted from the agenda. That's not the point. However, if we have to deal with issues which appear to have those characteristics, we have to work very hard in order to get the audience 'hooked'. You have to have an angle on it (figure 1). Paul Caro in his excellent contribution (p219) gave several examples on how to build up stories, and how to find the connections between front-line research and everyday applications.

When presenting contemporary science, it is worthwhile starting by trying to identify the players. During this conference, we identified the following players:

- scientists and their institutions (universities, research institutes, scientific societies, scientific journals, etc.)
- 'money power' (private companies, government agencies, funders)
- the media (figure 2)
- educational institutions (including science museums and science centres)

The division between the players is not always clear. Scientific societies are not immune from having political or vested interests and fiscal power, as the debate surrounding the exhibition *Science in American Life* at the Smithsonian, elucidated by Arthur Molella, shows (p131). Personally, I think it is a wonderful exhibition, and that the debate it has stimulated actually proves its success. Dominique Leglu, from the French newspaper *Libération*, pointed out that museums and science centres should perhaps do more to show that, in the real world, new technology does not always run smoothly; it sometimes goes wrong. Indeed, for the journalist, the best news is when technology goes wrong.[6]

After identifying the key players, the next question should be: what is the optimal

Figure 1. Basket-ball playing rats are a popular feature in a short presentation at the Heureka science centre. This has proved to be a powerful way of introducing audiences to ethology, the science of animal behaviour.

Figure 2. The media do not always present contemporary science as accurately as most scientists would wish. It is crucial that museums and science centres maintain their reputation as providers of high-quality science.

division of labour? Theoretically, it should be based on skills and possibilities. I think it is fair to say that research institutes tend to be research orientated, although the papers by Dominique Cornuéjols (p107) and Heather Mayfield and Peggie Rimmer (p101) showed how research facilities are beginning to address seriously the issue of communicating their work to the public. 'Money power' tends to have specific economical and political interests. The media, on one hand, are there to make a profit, but on the other hand, they provide a service to the public. From this analysis, the only group that is clearly and solely audience- and public-service oriented, is

the educational institutions. Douma made the point that, for too long, science centres and museums have operated in the shadow of other groups' preoccupations: of scientists, of formal educators, of politicians, of technologists and of industrialists.

The public will expect a good service and a balanced view from science museums and science centres. We are expected to provide lucid explanations, using good audiovisual aids. We are probably good at making interactive exhibitions. Wolf Peter Fehlhammer said that the ultimate interaction was to have scientists available for the visitors 'to touch' (p41). Many other speakers talked along these lines: Melanie Quin on drama (p77), Munkith Al-Najjar on Science North's collaboration with the Sudbury Neutrino Observatory (p67), Edward Wagner on the Cutting Edge Gallery at the Franklin Institute (p155), and so on. At science museums and science centres, we can provide a forum for interaction with experts in an exhibition setting. In order to provide this service, we have to cooperate with the other key players. These issues were discussed in the contributions by Richard Piani,[7] AnnMarie Israelsson (p139), Gillian Thomas (p125) and Arthur Molella (p131).

The problems of presenting contemporary science are very much the same as those we encounter when dealing with controversy and conflict. How do science museums cope with scientific controversy in their exhibitions? The problems of contemporaneity can be articulated around four dimensions.

The prematurity problem, i.e. the situation in which evidence is not yet considered watertight by the scientific community. However, this situation is part of the scientific process itself. Most great truths start as heresies. Wolf Peter Fehlhammer considered this kind of risk-taking as inherent in the scientific process (p41). In his view, it should be shared by the science museums and science centres. David Lowenthal said that we just have to live with it; we have to describe the scientific process, explain why there is uncertainty and cope with it (p163). At one point, conference delegate Barry Mooreshead of the Exploratory, Bristol, raised the issue of whether presenting conflict risks public trust by questioning the reliability of science. This needs to be considered carefully at a time when some cultural commentators are pointing to the emergence of an 'anti-science' movement.[8]

I can identify three issues. *The complexity issue*—the situation when understanding the issues at hand requires so much specific knowledge that the majority of the public is excluded from the debate. To my mind it was curious, and perhaps a source for optimism, that many speakers did not recognise this to be a problem. *The dimension of critique vs fairness* is that science museums and science centres cannot take sides while a discussion is going on. We should give a fair representation of all valid arguments (acknowledging the difficulty in determining validity in a scientific debate). There should also be a fair presentation of vested interest. Molella provides a good example of the issues that may be involved (p131). *The problem of context* or, as Schaffer pointed out, positioning contemporary or current research in history, locating a specific piece of research in society, and dealing with the technological implications, the economical consequences and the social and ethical issues (p31). Philip Campbell[9] remarked on the conspicuous absence of commercial aspects in science museum and science centre exhibitions. The context is very much a question of storytelling; here again I refer to the contribution by Paul Caro (p219). He shows that there is a direct connection between everyday life and advanced technology: it is just a question of finding the connections and developing them into stories which will attract interest. For example, Ivo Janousek of the National Technical Museum in Prague, discussed linking art and technology through exhibits on industrial design.[10]

There have been many examples of the issues which are under debate: Simon Schaffer (p31) referred to the life-on-Mars question, and an entire section of the conference on biotechnology dealt with these questions. John Durant said that the question is not how to excite our visitors about biotechnology, but how we should present it (p235). Museums and science centres should be more courageous in the portrayal of biotechnology and the case studies presented here reveal some of the issues that we will face as more exhibitions on this complex and emotive subject are developed.

At the organisational level, science-museum and science-centre professionals seem to know how to go about solving these problems. A major conclusion that emerged from this conference is that the complexity of the issues involved requires interaction with leading scientists. In this context, one may ask whether we are able to depict truthfully the scientific process. If you want to describe why you have controversy and debate, you have to describe the whole process. This is an area in which science museums and science centres have perhaps not been particularly successful up to now.

A truthful picture includes presenting the hard labour involved in science. A classic example is the work on Salvarsan, a medication against syphilis, by Paul Ehrlich back in 1909. His 606th experiment was the successful one. Of course, there is much more to science than hard work. It is also intellectually fascinating and emotionally rewarding. Neither of these aspects is usually dealt with in our exhibitions.

One aspect of the contemporaneity issue is that it requires a fast response. Developing exhibits, and especially interactive exhibits, usually takes some time. The paper by Helena von Troil sheds some light on the issues involved (p249). New technologies, and the Internet in particular, may provide a solution. Many interesting discussions took place at the conference on the role of Internet. It offers great opportunities for presenting contemporary research, and I have the feeling that science centres and museums have perhaps not yet grasped the significance of this medium. The papers by Roland Jackson (p173), and Andrea Bandelli and James Bradburne (p181) point out the great opportunities. There are a lot of open questions, of course, but it is quite clear that the Internet applications in the museum field are going to grow enormously.

Museums also have other forums for fast reaction, such as lectures, events and public debates. Essentially this means: find a scientist, put him or her on the stage and let the public in. As an example, I use my own organisation, the science centre, Heureka, in Finland. When the debate on bovine spongiform encephalopathy flared up in Europe, we had one of the two scientists in Finland who were working on prions giving a public lecture at Heureka

within three weeks. The lecture got good press coverage, which served to inform the public further.

The question of contemporary collections was briefly touched upon during the conference. The general consensus was that objects provide a starting point for interpretation—for storytelling. The subject of science centres versus museums was briefly touched upon. I personally do not see any real philosophical conflict here. I tend to regard science centres simply as museums with immaterial collections.

Change is a continuous feature of science, and, indeed, of the world. Douma indicated that the value of intellectual property has dramatically increased. This, again, leads to an increased need for adult and informal education. Are science museums and science centres responding to these changes? We have always used the best available technologies. In this regard, there is perhaps more need to develop in terms of speed and quantity than to produce something qualitatively different. From this perspective, the Internet could actually be regarded as an extension of existing services. It applies new technology to things we are already doing and is more efficient in getting out to a wider audience than are more traditional methods.

Is the change big enough to produce a paradigm shift in science museums and science centres? The early twentieth-century institutions, such as the Deutsches Museum in Munich, the Science Museum in London, and the Palais de la Découverte in Paris, brought interactivity into the museum. The Exploratorium in San Francisco took us one step further, by bringing in real open-ended experiments. Is there a change big enough to justify calling it a generation leap or paradigm shift? I have talked about third-generation museums myself,[11] but I am not certain any more. However, the changes we see involve two things, and this has been made quite clear at this conference: science museums and science centres will become less dependent upon location, but more dependent on people.

Notes and references

1 Douma, J, presentation to the *Here and Now* conference (London: Science Museum, 21 November 1996)

2 Kuhn, T, *The Structure of Scientific Revolutions* (Chicago: University of Chicago Press, 1970)

3 Persson, P-E, 'Science centres: dedicated to inquiry and exploration', *Physics World*, 9 (1996), pp55–56

4 Falk, J and Dierking, L, *The Museum Experience* (Washington, DC: Whalesback Books, 1992)

5 Salmi, H, 'Science centre education. Motivation and learning in informal education', *Research Report 119* (Helsinki: University of Helsinki, Department of Teacher Education, 1993), pp1–202

6 Leglu, D, presentation to the *Here and Now* conference (London: Science Museum, 21 November 1996)

7 Piani, R, presentation to the *Here and Now* conference (London: Science Museum, 21 November 1996)

8 Holton, G, *Science and Anti-Science* (Cambridge, MA: Harvard University Press, 1993), pp144–89

9 Campbell, P, presentation to the *Here and Now* conference (London: Science Museum, 21 November 1996)

10 Janousek, I, presentation to the *Here and Now* conference (London: Science Museum, 21 November 1996)

11 Persson, P-E, pp55–56

Index

Endnotes are referenced if they provide additional textual information, and are given as 'n' qualified by the endnote number in parentheses. Italicised page numbers indicate photographs.